Simple Questions Complex Science

ISBN-13: 978-1539419303
ISBN-10: 1539419304
In Mono Colour

AKA

An open letter about

Xepted Scientific Newton Mistakes

ISBN 0-9584410-1-4

By PEET (P.S.J.) SCHUTTE
AS A PART OF
MATTER'S TIME IN SPACE
– THE THESIS –
ISBN 0-9584410-8-1

TRANSLATED BY
PEET SCHUTTE.

FROM THE ORIGINAL AFRIKAANS:
"MATERIE SE TYD IN RUIMTE"

WRITTEN BY PETRUS S. J. SCHUTTE

I. S. B. N. 0−620−27041-1

WRITTEN BY PEET SCHUTTE
©KOSMOLOGIESE EN ASTRONOMIESE TEGNIKA

TO WHOM IT MAY CONCERN,
I am Peet Schutte, the author of the above-mentioned book/s
PROVE ME TO BE INCORRECT IN ANYTHING I SAY!

To whom it may concern and all others reading this document:
This is my introduction and this is my prologue:
But before I can commence with that task I have another duty administer: I AM ABOUT TO WARN EVERY PERSON IN SIGHT OF MY WORK ABOUT MY SLENDER ABILITIES.

Therefore in the light of what the most respected academic group on Earth accuses me of, I therefore have to issue a most serious warning to any person with the intention of making some kind of inquiry to the content this book holds, then the most concerning matter involving any content within the pages of this book you hold are that you must please seriously consider that where the stating declares the possibility that the content in this book has been (written by...) then don't take the announcing Written By Peet Schutte (Petrus S. J. Schutte) very seriously for there are grievous doubts leaving considerable dispute about the possibility, which underwrites the authenticity of Peet Schutte achieving the (written by...Peet Schutte) status.

Please take note of the following dehortation.

In the light of the reference to me serving in the capacity as being responsible for authoring, (written by...) in line of keeping fairness and justice to members of society, where all civil beings should carry reputed honesty, then: **Please be warned before any** reader starts reading about the following extremely serious admonition: I am bound by my conscience to warn all intended readers that I am placed under caution by the Academics in Physics.

Those most esteemed members responsible for the guardianship and maintaining the ethos in physics are of the opinion that I, Peet Schutte, am unable to write any book on the science of Physics as well as Astrophysics. Therefore, I, Peet Schutte, must declare that I should be considered as not very able to write anything, because I am incapable thereof. I suppose, I merely generate new information, which I establish as thoughts and then gather as concepts.

I further collect the result as words, which I put on paper using alphabetic symbols. I then compile that in a format that others may confuse with a book, but a book it cannot be, since the Masters in science found me unable to write a book.

But before you go further and follow my arguments, I first have to level with you about how academics view me in the position I hold. Please do not allow me to fool you, for this then cannot be, or represent a book. Now I have done my duty in warning everyone and in that, I denounce further participating with any purposive intention to wilfully bring down the crux of civilization by acting unacceptable and irresponsible.

I didn't write a book since I am not schooled to do so. It is my guess that I merely generated uninformed thoughts, which I collected as alphabetic symbols and plotted that in ink on paper. This effort I achieved from harbouring my delusional ideas spawned by a dehumanised brain. It only proves my weak and under developed mentality, due to my lack of an informed insight that is a typical symptom that all those have that is suffering from a disadvantaged past that one can only have when the person obviously lacks formal education.

While you are reading the letter deciding to regard or dismiss my work, then also please keep in mind when reading my language used and also please give credit where it belongs...if you do find linguistically improper use of words or misspelling, then remember that I am a feeble minded motor mechanic and not a literal giant.

I do find much pride in my status as being Afrikaner and would like to have my names used by pronouncing it in the manner Afrikaans dictates...therefore I would sincerely appreciate the courtesy when readers will take note that my name and last name are pronounced in Afrikaans, which is originally from Dutch and must be pronounced that way. Peet one would pronounce "here" which is the closest English to the pronouncing of the "ee". The "Sch" in Schutte is pronounced exactly as school is where both actually are pronounced Skutte or "skool". By pronouncing my name in Afrikaans you do me the utmost courtesy any one can. Being an Afrikaner is what I am most proud of.

Should any person challenge me about the legitimacy of the statements and content of this book, please do so at any time after you have familiarised yourself with what it is I say. However do not do it on the grounds of only the information provided in this book.

For such persons believing totally in the accuracy of science, the believability of Newton, and the uncorrupted nature of Physics academics first get to know the truth by going to and reading www.sirnewtonsfraud.com, www.questionablescience.net as well as http://www.singularityrelavancy.com/ and see what facts you ace.

There are so much facts pointing the truth and that much more detail when you read the six part theses called **THESIS"**

As you read the title of the book
www.SIRNEWTONSFRUAD.com

I know and realise that you are disgusted by my attitude when I degrade the name on which physics are founded. In this introduction part I am going to show you just some of the deceptions all students are forced to believe since all physics students are forced to believe in Newton, Sir Isaac Newton that is.

In the following am giving you a choice. You can say I am going to commit fraud by aligning the planets' positions according to mass but then Newton has committed the fraud because I only follow his lead. If I am judged to be the culprit that is guilty of deception then it is because Newton misled me.

You can choose.

You are expected to believe the following: Newton stated under the nametag of Kepler that there is so called Conversions for "Unknown" factors.

My personal advice to any potential reader of my writing: My style needs very careful reading: Do not try to speed read or do glance reading but read every word I use carefully and you will do well in understanding because with my writing style it requires much concentration. I don't waste words and I compress factual information.

What does A CONSPIRACY to UPHOLD a SCAM means and how would those most- honourable gentlemen with much integrity conducting science benefit from being dishonest about science I hear you ask in amazement and surprise?

There is no factor such as mass in the Universe. There is no evidence of a factor such as mass that holds any validity throughout the Universe. There is no proof that the Universe indicates the presence of gravity by the measure of mass forming a pulling power and while science conducts an entire religiosity based on this falsified belief, any such notion is falsified truth. **Using science based on the idea that there is a pulling force such as mass forming gravity is as valid as giving Snow White seven dwarfs and then beginning a religion on that basis.** There is a factor such as weight but there is no pulling of anything towards anything by magical forces forming gravity or whatever. **There is a conspiracy of conducting fraud by claiming non- existing forces but such claims are utterly fraudulent.** I have been trying for twelve years to introduce the true forming of gravity but all Physicists I have encountered prevented me of doing so. They stop me because my work makes Newtonian science and when removing the notion that a pulling force of gravity works by the value of mass, most of their work becomes science fiction that falls apart in substance. Read this and see how **students in physics** are methodically **brainwashed** to get the students to believe in the absolute accuracy of science. Professors and teachers participate knowingly or unknowingly in this thought manipulation process by means of conducting **mind control.** By applying this **mind-altering process** those teaching physics ensure they subdue students into becoming mind-altered zombies. It is a process going on for centuries and which without science would have no foot to stand on in the modern environment. By presenting incorrect, falsified or unproven facts and other untruths as proven truth they exert **thought control** and thereby change the student's ability to appreciate what is correct and believable logic and then force students to discard such judgement ability in favour of accepting the institutionalised untested norms and values of science in order to unequivocally believe in science. The accepted teaching methods force students to comply by compromising their better judgment and then systematically to capitulate under teacher pressure by making their own what science prescribes what should be believed. I prove this and you get this for free so what have you got to lose…**but you can get wise to what forms a better understanding about science**! By using the building blocks that forms the Universe I take you back to the instant the Universe started and I show you how the Universe fits like a jigsaw puzzle.

If you think my accusations are baseless or the ravings of a madman then go on and download what you have opened and read for yourself. What you download is free and I do not benefit financially from this explanation I present to you.

WHOM IT MAY CONCERN,

I am P.S.J Schutte, nicknamed Peet. Being a white South African my mother tongue is Afrikaans and my second language is English. I have per suiting a new cosmic theory that I partly present in a six part theses, of which the investigating research began in 1977. First I located what was wrong in physics. I compiled my presentation of The theses called The Absolute Relevancy of Singularity and then six separate thesis parts forming the theses published through LULU.com which I saw as way the only manner whereby I could generate funding by which I would be able to have the twenty seven books I already wrote linguistically edited and then to have the books published on a Print-On-Demand basis. I compiled a new cosmic theory by which I eliminated all the incorrectness that Newton has burdened science with but with this being my opinion I did not find a garage full of academics supporters waiting to applaud me and to uphold my views on the matter. Yet still I was not going to be ambushed by their relentless stonewalling my efforts and blocking my efforts in introducing both the incorrectness and the new cosmic theorem I concluded. Their blocking convinced me about a Conspiracy in Science in Progress and this spurred me on to tell the entire world about their brainwashing of the minds of students.

This kept me busy for the past going on to twelve years on full time basis whereby I was trying to introduce my findings to many academics without finding much joy from my efforts.

This past eleven years plus saw me go without any income as I tried to get my theorem recognised as well as get my warning noted. Going without a steady income left me almost destitute and in order to find a manner to get my theory across to the attention of influential readers, I decided to publish a theses of six books electronically as to try and get around the stranglehold of Newtonian bias controlling science at present worldwide. I decided to publish electronically which those in power do not control.

With my first language not English and the books not linguistically checked by an expert there are bound to be language errors that readers will notice. In the past I tried to check my work myself but after checking say one hundred and fifty pages for language corrections, then after days of toiling instead of having corrected work I ended having four hundred pages of newly written information which is still not linguistically corrected but holds a lot more information.

The language and spelling errors compiled instead of reduced. This is because my priorities lie elsewhere. I aim to spend money on correcting the work as far as language goes, as I receive money in the selling of my theses and in the hope that I will receive money. I will have all my work including the one you are reading edited professionally and corrected as I find money to do so...But first I have to get the public aware of the problem to get the academics to appreciate the problem.

www.singularityrelevancy

from Lulu free of charge

email: **mailto:info@questionablescience.**
order@sirnewtonsfraud.com

I do find much pride in my status as being Afrikaner and would like to have my names used by pronouncing it in the manner Afrikaans dictates...therefore I would sincerely appreciate the courtesy when readers will take note that my name and last name are pronounced in Afrikaans, which is originally from Dutch and must be pronounced that way. Peet one would pronounce "here" which is the closest English to the pronouncing of the "ee". The "Sch" in Schutte is pronounced exactly as school is where both actually are pronounced Skutte or "skool". By pronouncing my name in Afrikaans you do me the utmost courtesy any one can. Being an Afrikaner is what I am most proud of. If you have been a scholar at school you would know how to pronounce Schutte.

I always submit articles to well known physics magazines but my articles are rejected on the most unappeasable grounds and for the most outrageously ridiculous reasons the Newtonians can think of. I explain how gravity forms but I am rejected because they are of the opinion that my work does not meet an acceptable level of standard since I am at odds with the way science in the present think about gravity.

I say and I prove there is no such a thing as "mass" with the ability to "pull" anything. I do not say a person does not have weight but I say no person has "mass" that brings about a force that pulls anything closer by using gravity.

According to the Big Bang principle everything is growing and moving "further apart" and therefore there is no "mass" pulling anything closer according to "body mass" because nothing has "body mass". But my views are rejected because my views clash with the wisdom of **the Masters forming the principles that direct the thinking of science in science according to science**. If you want to read how corrupt the thinking is in science then read on but do so with the sole purpose to prove me wrong. That is one thing not one of the **Masters forming the principles that direct the thinking of science in science according to science** carrying all the wisdom they can manage could ever do!

In the books on offer through this web page and in which I am introducing a totally new concept in terms of gravity, the proof I bring is true about gravity being formed as a result of these phenomena. In the past science hardly recognised the existence of such phenomena although they are known to science for centuries.

They are known as

1) The Lagrangian system
2) The Roche limit
3) The Titius Bode law
4) The Coanda affect

However, since the explanations that I provide holds a completely new line of thought, there are just too many and too numerous wide ranging facts behind that which forms the complete picture as a whole, this leaves me unable to include a full introduction in a space as small as that which a web page will allow. The explaining of such a totally new approach includes for instance those phenomena science this far failed to understand and which I have named as the four cosmic pillars. With these facts being altogether new to science, I find academics showing very little willingness to consider the acceptable value thereof. I recon it must be the result of science seeing so many idle explanations in the past and then proving to be senseless as much as being little impressive, therefore my mentioning it without bringing and substantiating proof will be fruitless and counter productive.

I found the manner in which to interpret Kepler's formula as $a^3 = kT^2$ and I found that when dealing with Kepler's formula, we should not see a^3 as space but we should see singularity being positioned in space in relation to singularity forming relevancies. What brought the answers was putting singularity in context with Π. Doing that placed me in the position to discover what gravity is and how gravity operates to form the Universe. I saw that Kepler's formula should instead use Π. By placing Π in relation to gravity I manage to find an explanation for the four cosmic phenomena. Everything that has anything to do with gravity forms a circle albeit that it is called the curvature of space-time or gravity bending light or forming a round galactica, the connecting factor is gravity which implements Π. Gravity or another name used to call gravity would be time is running on the measure of Π and every aspect of cosmology integrates Π as the basic concept on which cosmology is founded.

Because my views do not echo the commendable praise attributed to the greatness by which Newton is commemorated, my work is purposely and very much wilfully poorly received in the world of physics and astrophysics and by that I find very little willingness in any understanding shown in the ranks of Newtonian science. This work contains ideas about the introducing of a totally new concept on explaining for the first time ever the working principles of gravity, a matter that eluded Newton no less. I decided to offer four books that introduce the explaining of these concepts in e-book format. This method of publishing rests totally on a financial basis. I tried to introduce the four phenomena as a concept by using a web page but found such introduction is far too comprehensive in having just too many and numerously wide ranging facts that form the complete picture as a whole to be comprehensibly appreciable, and therefore on account of that realisation that I was unable to include a full introduction in a space as small as that which a web page will allow, it gave me the idea of introducing this new concept via electronic publishing. As my other books I sell by printing are all hundreds to thousands of Mega Bites of information, I had to revise the layout where each is to have fewer than twenty mega bites. This motivated me to only introduce the concepts in producing small books that then could be sold via the electronic publishing media as to allow persons to first acquaint themselves about the viability of the concepts and the feasibility of this new approach I introduce. If any person shows interest in finding out more about any of the books, please click on the book of interest and discover something in science no one yet has ever heard about.

The main issue of finding the value and the meaning of the four phenomena was to connect gravity to Π. Gravity is much closer connected and is much more intimately related to Π than it ever can be linked with mass. By giving each of the phenomena a measured value in terms of Π solved every riddle connected to the phenomena and not only did the phenomena become purposely clear but also the working principle gravity…

I prove that gravity is Π and the four cosmic phenomena forms gravity by producing Π when objects move in relation to each other.

By using the above **the four cosmic pillars,** it enable me to **present the proof** where I now can explain what conditions bring on the **sound barrier**. By proving it **is gravity that** the individual structure generates motion above and beyond the gravity the Earth provide is what is producing individual **motion** that the independent object earned within **the sphere of motion that the Earth's gravity provides** where the **independent and individual motion** put the relevance that gravity has beyond the conserving means gravity has **where** the space **that is serving the** independent object **is** independently in motion. **The** adding **to the** independence **on top of the** normal structural independence **is creating** more individualism **by the** independent motion **of the** individual structure **being** apart from **the motion that the** gravity of the Earth **provides.**

1) **I found the** location, position **of** singularity $\Pi^o\Pi$ **as a factor forming space-time**

2) **I found** space-time **by dissecting Kepler's formula in relation to** valuing singularity **as** Π^o

3) **I not only found but I also** proved space-time **by** showing how movement **forms** Π.

4) **I found that the** working principals **behind** gravity **as a cosmic occurrence are forming** Π.

5) **I found the reason for the** Roche limit **and explaining the resulting of** that in relation to Π.

6) **I found** out why the Lagrangian system, **becomes** Π **as** the building form **of the Universe.**

7) **I found why the** Titius Bode law **mathematically provides the** foundation of Π.

By proving that the Coanda affect is gravity assisted by the other three phenomena and thereby through activating singularity it formed space-time, I was able to link gravity with singularity in applying the measured value of what forms Π and then I could link singularity to gravity by the forming of the value of Π as gravity. That I managed by seeing that it is the way that Π forms that gravity also forms.

The fact every one misses is that any structure that is not part of the earth crust in the structure has own gravity that is stronger than that of the Earth which provides the independent individuality the structure with unique structural space has. The gravity of the Earth strive to incorporate all in the sphere the Earth has into the structure the Earth has and therefore the fact that the structure that forms a factor Newton called mass is not incorporated as yet, shows defiance and individuality in the first place.

Also read

A Conspiracy to Commit Fraud on a Cosmic Scale

Part of

www.sirnewtonsfraud.com

WRITTEN BY PETRUS S. J. SCHUTTE

ISBN 978-1-920430-07-8

©KOSMOLOGIESE EN ASTRONOMIESE TEGNIKA
All rights are reserved.
No part, parts or the entirety of this book may be reproduced by publishing, electronically copied, duplicated by whatever means that form reproduction or duplication, without the prior written consent of the copy rite owner.

This is a pre-runner of and going under the Title heading of

A Conspiracy in Science in Progress

Part of www.questionable science.com

http://www.lulu.com/content/e-book/wwwsirnewtonsfraudcom-part-2/8132511]

We all are as clever as our stupidity will allow us to be...
Therefore we are not in a fight to get wise
But our life-long fight is to destroy our stupidity
And rise above our shortcomings.
We mist stop looking and the little we know and concentrate on the vastness of what we do not know
This is your introduction to physics and every aspect that brings about information to physics.
Physics is not mathematical equating of more rubbish. It is ideas mathematics can never prove.

Have you ever stood outside and thought how is it possible to get that wide area of space that you see into something as small as your eye? If you have never thought of this you have never though about the cosmos because this is physics and the start of physics. That is how one starts to understand the cosmos and on this thought I base everything I introduce as physics. You are about to experience ideas you have never even thought about and I am introducing to you concept that you are unaware of that it exists.

With my first language not English and the books not linguistically checked by an expert there are bound to be language errors that readers will notice. In the past I tried to check my work myself but after checking say one hundred and fifty pages for language corrections, then after days of toiling instead of having corrected work I ended having four hundred pages of newly written information which is still not linguistically corrected but holds a lot more information. I prove gravity as component of science in a manner that was never proven yet and this proof changes all aspects of science and I am not exaggerating in the least. Every time I run through a check I realise I better explain something by adding some other factors just to bring more comprehension to subjects that is very difficult to grasp when you hear about the facts for the first time. The language and spelling errors compiled instead of reduced. This is because my priorities lie elsewhere. I aim to spend money on correcting the work as far as language goes, as I receive money in the selling of my theses and in the hope that I will receive money. I will have all my work including the one you are reading edited professionally and corrected as I find money to do so…But first I have to get the public aware of the problem to get the academics to appreciate the problem…but after you read this letter you'll know why they have an attitude.

What I am about to explain is not how the cosmos started because that is immeasurably more complex than the depth I am going into with this conversation. My intensions are to show why singularity is so important, why seven is connected to the birth of the Universe and how ten became the liquid of space. Looking at the night sky you see many flickering dots and spots and light coming from afar. The one star you see seems to be a near visible dot in the picture while the dot might be hundreds or might even be many thousands of times the size of the sun…and we think of the sun as big.

The dot is then that much bigger than the sun because the star we think we see could be a galactica hundreds of times the size of our Milky Way galactica but that shows in the sky as one little dot and yet that entire structure as big as it is many times our Milky Way, fit into our eye socket. But that is not all…there are trillions of such light images and they all fit into one eye socket. What we see is immeasurable and yet we see it effortlessly in the space our eye holds…how can that be? It is because we see dots and spots that form a Universe.

I have asked this many times before but I ask again: Have you, the person reading this, ever thought how it is possible to see that much information that you see at night when looking at the sky with only your eyes? Ever thought about how you are able to see when you see everything in the night sky and how that much light information can fit into such a small space as your eye? Have you ever sat back and think what vastness in information it is that you see when you see the entirety of the Universe when looking at the Universe at night and what the size is of everything of that which you are able to see?

But that is not all…there are trillions of such light images and they all fit into one eye socket. What we see is immeasurable and yet we see it effortlessly in the space our eye holds…how can that be? How is it possible to fit what we see into the space of our eyes we have? Think how much is the entire information that is visible at night and think about how all of that fit into the space your eye holds? This is what the Universe is and that is what the Universe represents and when I put that into a mathematical equation I wish to see those persons designing space whirls to calculate the volumetric space reduction with physics. Consider how big is what is visible and put that space into the size of what your eye can hold and ask your mathematically educated Professor in physics to find some ratio between what you observe and the size of your eye.

The ratio is astonishing, but more-over what is truly astonishing is the arrogance of man to think of his position, as being important while the space man holds is beyond any comparison in ratio to everything we see in the Universe we see. Think how small we are when we are able to see the entirety out there! Even if there was other life out there, what is the worth of it in comparison to what there is that we see? Tell your physics professor not to reinvent the Universe will his brilliant mathematics, but just too mathematically formulate how all the light forming a Universe, the visible Universe can fit into my eye.

No less telling is the next thought I present you with. Why would all the light coming from all over and everywhere meet you in the precise location you hold? The light left its place of origin and travelled for in some cases 12 billion years or more at the speed of light no less to meet you in the position you are at the location you are and all the light coming from everywhere possible comes directly to you. Makes you feel important does it not? What a thought this is for the further boosting of the ego that man cultivates!

It is not what science declares that is important but it is always what scientists don't declare that holds prominence and more so the reason why science keeps a silence about the information they do not disclose. It is never about what they say but it is why they don't say other things they keep quiet about. You will read how they never disclose the entire truth because science is about promoting one-sided and selectively opinionated information forming fraud no less. I have been per suiting a new cosmic theory that I partly present in an eight part theses, of which the investigating research began in 1977. In 1999 I compiled my theory and searched for a publisher.

I use Kepler's formula but not as Newton did when he raped the living daylights out of it because he had no idea what it was about. By using Kepler's tables correctly I show where to locate the very first spot that forms the point where the Universe started. The question about the Universe is how can whatever is in view, come stored as a parcel in an electron, and tell the entire story about the entirety out there locking all that data into the space of a photon I use to see things. That is physics and however you may try, not one person searched for the ability to calculate that part, although the cosmos gave Kepler the formula.

The United States Government brought in laws to cover the wide spread corruption that persons such as the founder of wikileaks reveal. They made laws to protect lies from surfacing and anyone that uncovers corruption in their ranks is jailed for uncovering the corruption. The media in every form is overflowing with all sorts of corruption going on in commercial banking, in politics, in corporate pharmaceutical medicine, in hospitals and every other industry... and even reasons why the West attack Arab states on behalf of other Arab states. Even the TV media is so one sided nothing they report is believable because behind every action we see the dark hand of Money Moguls brainwashing persons to believe the politicians' actions for going to war is morally justified! Then what about science? Why is science that Lilly white they make it out to be ... or is it? This is the whistle blower on fraudulent practising going on in science. This website is to inform about shocking detail everyone misses and reveals how deep the misconceptions are in science. This exposes details of blatant mind bending brainwashing practises science are committing to control the thoughts of unsuspecting students. In this book there is also a business offer that can be very lucrative for those willing in participating to help publish the uncovering of dark secrets in science that large money hides to keep the world slumbering in fogginess and twilight. Reading this book will inform any reader on any level more than any other book did before and reveal a financial offer that will not repeat again. You can take part in the battle against science fraud and be financially rewarded.

The media in every form is overflowing with all sorts of corruption going on in commercial banking, in politics, in corporate pharmaceutical medicine, in hospitals and every other industry...Then what about science? Why is science that Lilly white they make it to be or is it? This book is the **whistle blower** on fraudulent practising going on in science. This book informs about shocking detail everyone misses and reveals how deep the misconceptions are while this book exposes details of blatant mind bending brainwashing practises science are committing to unsuspecting students. In this book there is a business offer that can be very lucrative in participating to uncover dark secrets that large money hides to keep the world slumbering in fogginess and twilight.

Please read first: I write about cosmology, which is the science of the Universe and moreover stars and other cosmic materials. I explain for instance why there are stars amongst which are named neutron stars and Black Holes and I explain why there are stars and galactica and nubile and Super nova stars and why these cosmic conditions are what they are. I found the reason why the cosmic phenomena used by the cosmos are in place and what gravity really is. I discovered this for the first time in human history. Due to research I found the way the cosmos developed going back before the era called the Big Bang. In order to accept my work science has to reject critical aspects of science they at present accept unreservedly. It is about how and why time forms space to develop structures in space that hold materials …in short it is how stars and galaxies form and why stars sometimes don't form or destruct in violence.

This topic is not a fictional love story or the normal other general topics dealings with the some everyday commodity, which persons normally write about and in that instance then endures fashion trends or likes, and dislikes. My work is very much researched and has never been published and contains work that finds interest in a wide spectrum of persons. The facts I use in all the books are as about actual events as live as it comes as anything one would encounter on the news everyday and it generates a wide spectrum of interest over a wide area of all populations all over the world. The topic I elaborate on will be valid reading material as it is now applying, as it will be in a hundred years. It will not go out of fashion or be old news.

However when introducing the information the understanding level can become as complicated as much as only where a very few can understand or it can be as easy as all other normal reading material would be where the range of this wide spectrum of understanding the work depends on the proof given to substantiate the topic. The proof it carries complicates the level of understanding required.

Informing information about the series informing on Science Corruption
I have been doing science research about the Universe for many decades and I uncovered how the Universe forms space as it grows. The process is well known to science and has many names but up to now no one in science ever understood how the principles apply. I do not wish to venture much into this detail since it is lengthy to explain and covers many details but should you wish to know more it is as such in the books I sent to your institution.

The lack of understanding about how these principles function can only be contributed to mistakes in science about science at this moment and facts science ignore and these mistakes are what science are unwilling to admit. These mistakes academics embraces in the theories they apply and accepting these details would be equal to their denouncing of the correctness of every thesis written on cosmology. I have found irregularities in science and I reveal it but not only that; I give the correct approach as well and where necessary I supply the proof thereof. My concept changes science that those in science are unwilling to let their ideas become discarded in favour of accepting mine when they embrace my work. Since I find no acceptation or any endorsing about my work I now bring my work to the general public. Those in science frustrate me by blocking me. I wish to reach all persons reading on a level as normal persons do read to understand concepts so that I can introduce the approach to science I take. In this detail come the levels by which I have to do the planned introductions. I explain this somewhat tedious because I wish you to know what theme your firm will promote should we share a future in this venture.

In order too reach as wide an audience as possible the ideal approach is that I need to publish four books where the exercise will be ranging from introducing science in as simple terms as possible to silencing all academic critics completely. However the work containing the proof by which I address those I wish to silence is very complex and is not appreciated by everybody in the detail it deserves.
In achieving my goal I wish to publish 4 books

The exercise will be complicated should I not publish the four books at the same time because in the past critics ravaged my findings and in other books where I prove my findings very little to no one at all understood my work. To give you an idea: I prove Einstein wrong about light travelling across space within the realms of gravity. Explaining the concept is simple but proving it requires much more skilled understanding. Explaining the idea is simple because a straight line, a half circle and a triangle are all the same mathematical value of $180°$.but proving why the three forms are equal takes mathematics back to where it started and that very little number of persons can fathom. So I can explain it in simple terms but proving takes a long time and then I have to have evidence ready to silence those trying to belittle me where they have to purchase the expensive work to see how I prove it and understand the proof I present.

In every book information is shared but detail splits the content. If you read one you read the other partially because as they all are the same every one is also much different. Not one person has the identical same intellect. I wish to get my message across to any person that enjoys science but since we all have different fields of development we can handle information differently. I do not want to steal your money by letting you purchase at random. Not one person on earth knows you as well as you know your individual self. Sometimes people might think you are the village idiot while you know you are stringing them along and you are highly intellectual but you can convey yourself verbally not as well as the rest of people.

Even if you think you are a Master holding a few doctoral degrees, do not be a fool because what you are going to read will put you in the category next to the novice, The categorising depends on how you can manage NEW information and not what you think you know. The information I reveal is the result of a study conducted since 1977 and I can't let that go cheap because it is my life's work. It has never been printed and I spent decades formulating the formulas.

Peet Schutte

Another part of A SCIENCE CONSPIRACY

You are About to Uncover A Science Conspiracy

© KOSMOLOGIESE EN ASTRONOMIESE TEGNIKA

www.questionablescience.net

WRITTEN BY PETRUS S. J. SCHUTTE (Peet Schutte)
©KOSMOLOGIESE EN ASTRONOMIESE TEGNIKA
All rights are reserved.
No part, parts or the entirety of this book may be reproduced by publishing, electronically copied, duplicated by whatever means that form reproduction or duplication, without the prior written consent of the copy rite owner.

If you ask what is the difference between how I see gravity works and how Newton's gravity work? Newton says objects pull while I say space compresses thereby collapses by getting reduced and everything in that space condenses. I say space reduces and Newton said objects in space pull each other. While I prove nature, Newtonian science cheats, corrupts and manipulate nature to make science work in ways nature doesn't work. I say there is no pulling but it is space that compresses by objects producing material movement in rotating of or as gravity. The difference

between my approach and Newtonian's is one Universe away from each other. I show a functional Universe and Newton show mysteries of science. I prove everything that Newtonian science this far couldn't. I prove the Universe applies four keys by which gravity works instead of unexplainable magical forces. I show how nature works with the 4 keys while I show how science falsifies facts to make science seems to work by magical forces pulling. I bring you facts about what is true in nature and not what you think is true or science thinks is true but what nature uses as the truth. When you argue about what I say you argue with nature and what I bring you then you also falsify nature in favour of Newton as science do.

<u>I repeat this again to stress the thoughts with which I started but this is how I now start</u>: My introduction as well as introducing the readers to general cosmology has to be in a very brief and in a compressed manner but first, I have to give the emphatic warning to all prospective contemplating readers. I didn't make up those ideas I attribute to mainstream science as I went along but it is Newtonian science that cling onto the black magic Newton believed in. I don't wish to explain what science regard as the truth because I have too much of my own that is correct to introduce and to explain.

Should any person have any doubt about my statements concerning the official views mainstream physics have notwithstanding how ridiculous some might seem to be then go to the Internet and confirm that what I say about science is true and what I say about the incorrectness of science is true? This

picture personifies Newtonian science and I just have to use it twice to bring him the excellent example it says in portraying the Newtonian approach to astronomy. It is the same as lets pretend Newtonian science calculates and according to what they find they see there are mammals feeding calves milk while swimming in the see.

They find mammals are in the sea and mammals feed calves milk while swimming in the sea. They know cows are mammals and that cows are mammals that feed calves milk.

The next thing is they come on the six o'clock TV news and declare on international TV stations they found cows swimming in the see. They make a big splash and a hell of shouting a real cowboy hoo-hah about the most outrageous cows swimming in open water and the cows can swim on their tails while clapping their front hooves.

Now every cow farmer and milk drinker has to get excited because excited people part with money. The louder they shout the most outrageous nonsense the better-informed scientists they seem to become.

Their shouting on TV making the find as Hollywood glamorous as possible has the sole purpose to get everyone so exited that people would phone their congress man to push and urge him to give astronomic research even more money so as to try and find out what type of cow it is because there is a variety of cows that can swim in the sea. No one would dare challenge them about the cows not being able to swim in the sea because no one but they know it is not a cow but it is a dolphin and that part they leave very much alone because they have to be the experts that knows everything anyone could ever know! Are there cows swimming in the sea…that part he or she will never correct because then everyone will know he or she has no idea what swims in the sea. He or she will leave you believing mammals swim in the sea holding you under the impression that cows are mammals and calves feed on milk in the sea. That is swindling the truth.

They swindling by giving a list of half-truths because only science, Newton and God in that order is never wrong and that is religion. Lately someone was found to be wrong and science decided it had to be God that misplaced dark matter in obscure places in order to have the Universe expand instead of contract as Newton said it does. Science decide on behalf of God that God was wrong when the Universe expanded instead of contracted as Newton said it does because Newton couldn't be wrong so it's God that wrong. The way science present a star such as the sun is similar to a furnish burning to boil water. The furnace will burn until the coal is finished and then the star will **_die._** This view is carried over from since a time before they new about electricity or internal combustion engines or air-conditioning or nuclear energy.
I wish to make one fact very clear. I base my work on formulating the working process of four cosmic principles in Nature. These are:
1)   The Coanda effect
2)   The Titius Bode law
3)   The Roche limit
4)   The Lagrangian points.
I did not discover these phenomena because science knows about these phenomena for a very long time and in some cases even for hundreds of years. Science knows they apply and where they apply. When science discovered or allocated missing planets they used the law applying such as the Titius Bode law from which they deducted positions that they knew in that circle according to the planetary layout that the law predicts there had to be a planet according to the law. Science did not apply Newton's formula to discover and locate planets but they applied these phenomena and especially applied the law of planetary allocation to discover the precise location the planets discovered after Galileo.

Every one in science knows these phenomena is there and is in place and they rule the orbit set-up of the planets. The solar system functions according to them. These four laws on planetary motion that is used by nature at this moment and has been in place since time began are what apply and they dismiss Newton. If you argue with me about Newton being correct you better take your case to God or the solar system because the four cosmic phenomena is working in nature and nothing Newton said is applying in nature. This is a truth and a fact and a foregone conclusion and can never to be in doubt.

Brainwashed as you may be in believing Newton you can't either side with my view or decide on Newton because it is not a case of choosing between Newton and me. I'm out of the picture! It is either telling nature to listen to Newton and change what is in place or read and see what is in place and what is applying in nature all along. The phenomena are what we find to be used in the cosmos while Newton is in the imagination part of the minds of scientists and nowhere else. If you don't believe me and if you wish to discredit me first find out a little more about science. Then deflate your ego as to what you think you know.

Science never mentions these phenomena because science can't use Newton and explain these phenomena or use these phenomena to prove Newton. These four phenomena that the cosmos uses as we speak have been in place ever since the Universe formed. Since science can't explain the phenomena and the phenomena destroy the credibility of Newton science avoids these phenomena as if it brought the plague. You can't choose between Newton and me because I did not put the cosmic phenomena in place. All I did was doing a study since 1977 to formulate how and why these phenomena work and how these phenomena keep the cosmos and the solar system working. I am the first in history to show why they work. We are all been brainwashed for centuries to believe Newton. Should your brainwashing kick in and you have an axe to grind with me about what I say, then first prove these phenomena are not in the cosmos and are not applying to form the laws that the cosmos put in place as gravity. I only found out how they work and why they work and I did not make the phenomena work. All those clever stooges that have so much to say even before you read first learn what is in place before getting so opinionated.

I am not fighting science or the credibility of Newton or what might be true or not true but I am fighting centuries of brainwashing and I have to dismiss the brainwashing and the systematic mind control that those teaching science inflicted on us all. My fight is not about what is true or not true but what is accepted as culture and which was not even once been proven by science. Please do read on to investigate before becoming self-opinionated.

Please be warned about aspects of this book some may find offensive. Apparently in the following accusations aimed at scientists I use degrading language associated with persons of crude behaviour but the thoughts are not that petty. With this book I aim to address the public that are not well developed in all the facts just as I was a student but after decades of study I decided that it was time I placed these thoughts by pen on paper, I do this to try and find a means to reach the young minds with young thoughts and from humble and crude ideas I developed a new Cosmic Concept I wish to share. I wish the public to realise what science is hiding under a cover-up conspiracy. I thought it was time to lift the cover about information that I came to realise years ago when I started out thinking. I leave the language unchanged and I know too well that in the eyes of the Most Respected Physicists that the language, grammar phonetics I use is unsophisticated. To the Highly esteemed well-to-do Academics in Physics all this information I provide in these statements and concept also comes across a well below a standard, which those well-educated and highly developed Newtonians are used too, but also the way they frown on anything regarding me. I am by now use to their frowning upon me. Hey, for you members of the public I have books written that are coming to press, which they will not want you to read… This crudeness of my unwaveringly impoliteness is distasteful too the highly educated members of the physics establishment and my writing also lacks every bit of linguistic sophistication and terminology they use to avoid the truth.

I see what no one else saw in 500 years... do you believe that because I don't. I think I am just the only person or then the first one that felt the need to expose this culture of total corruption! I see misconduct that is rife in physics. I accomplished what I aimed to achieve by solving misguided concepts and that I produced as I intended when I started out finding a path amongst the maize of corrupted science by using arguments that Newtonians never even bothered before to give a thought. If the language is poor, to that I will admit but the ideas surpasses even the best Newtonian mind because where it ventures as crude as it is according to the uttermost sophisticated members of science, and yet where I am no Newtonian brain even realised the ability to go there.

The pictures seen below I explain the principles by proving information about what happens when nature performs because I managed to formulate the working process of the four cosmic principles or building blocks in Nature.

These are:
    1) The Coanda effect
    2) The Titius Bode law
    3) The Roche limit
    5)   The Lagrangian points.

As things stand mainstream science are unable to explain the cosmic principles applying and that is because science do not understand what nature uses as
    1) The Coanda effect
    2) The Titius Bode law
    3) The Roche limit
    4) The Lagrangian points.

There is an almost hundred percent chance that you never heard of these phenomena before notwithstanding that the entire Universe is built by these phenomena and these phenomena form the building blocks of the Universe.

The reason why you never heard of it and science never mention the importance of the phenomena is because science are unable to explain and so they rather ignore the importance than press the importance.

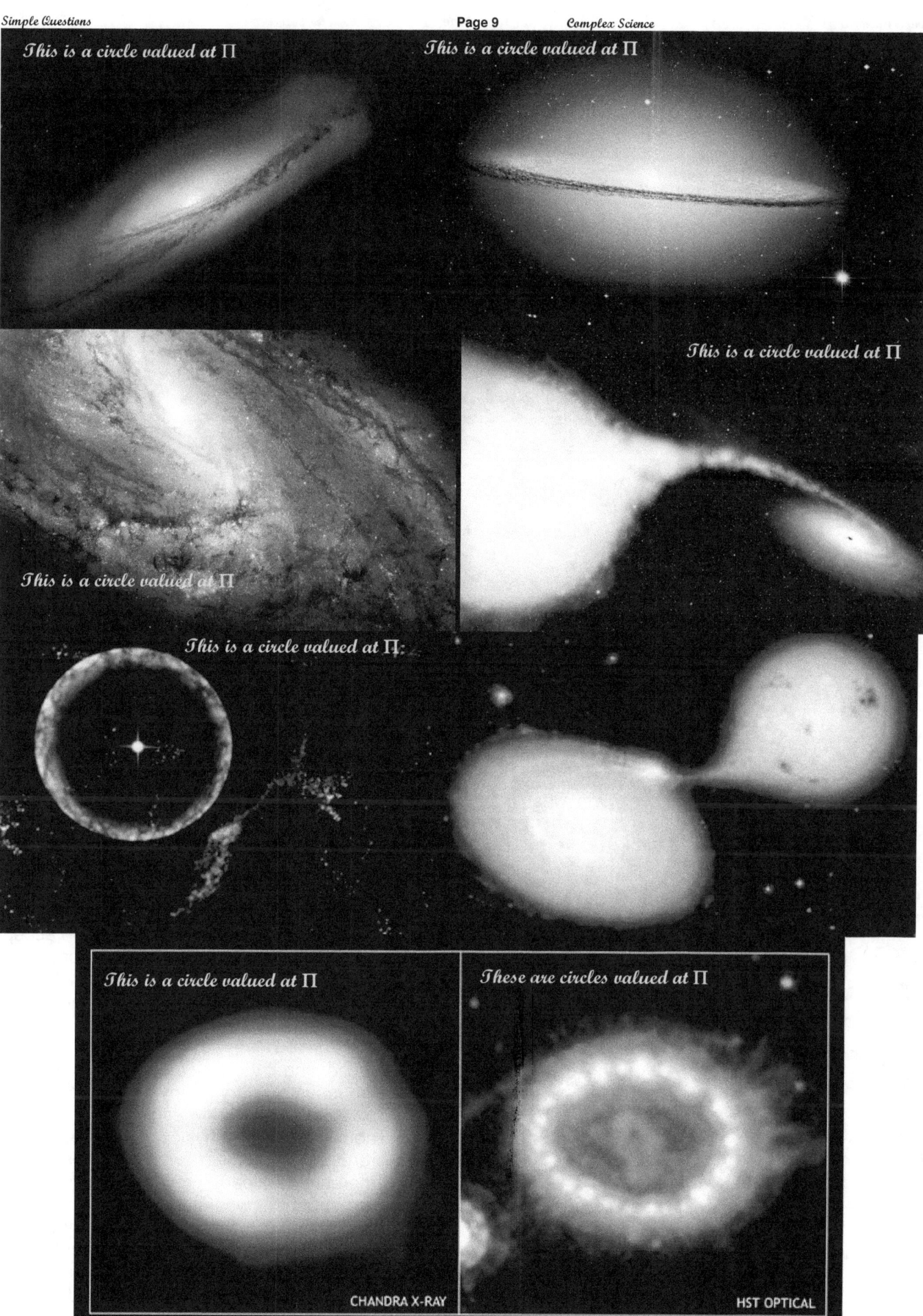

These are circles valued at Π

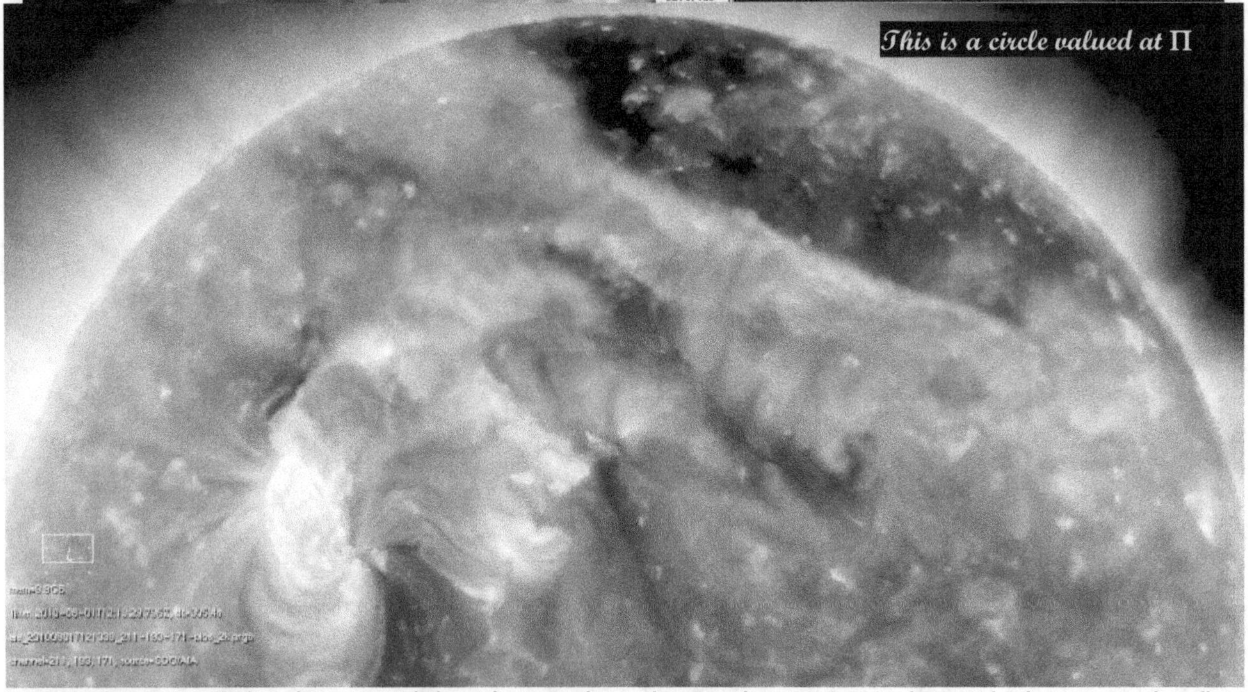

This is a circle valued at Π

These are circles valued at Π

Science at best can't begin to explain what makes these phenomena pictured above on the first few pages do occur or what happens when these phenomena occur. I can and I do explain in detail why these phenomena happen and I prove what I explain. Then why would science in its current concept not accept my explaining? If they do accept my explanation about why my concepts are correct then they have to rubbish their entirety of everything they think they know and trash every thesis down to the last dissertation they ever wrote on science presenting it the way they do. This will cost the industry every penny they spent this far on science. Acknowledging my concepts about cosmic physics and cosmology renders all their work useless and moreover it clearly proves how wrong they are about the woefully outdated Newtonian approach. It will become clear how little they understand about science because everything they say they understand is incorrect and applauding my view makes every book published on cosmology instant science fiction. You think they will embrace my work…hell no and now I turn to the public. I let the people decide what makes sense and what is rubbish. Purchase this book and then you decide! You will have to choose between nature and Newton.

Science at best can't begin to explain what makes these phenomena pictured above on the first few pages do occur or what happens when these phenomena occur. I can and I do explain in detail why these phenomena happen and I prove what I explain. Then why would science in its current concept not accept my explaining? If they do accept my explanation about why my concepts are correct then they have to rubbish their entirety of everything they think they know and trash every thesis down to the last dissertation they ever wrote on science presenting it the way thy do. This will cost the industry every penny they spent this far on science. Acknowledging my concepts about cosmic physics and cosmology renders all their work useless and moreover it clearly proves how wrong they are about the woefully outdated Newtonian approach. It will become clear how little they understand about science because everything they say they understand is incorrect and applauding my view makes every book published on cosmology instant science fiction. You think they will embrace my work...hell no and now I turn to the public. I let the people decide what makes sense and what is rubbish. Purchase this book and then you decide! You will have to choose between nature and Newton.

If you have never heard of these phenomena it is not that surprising because you either have to believe in Newtonian magic Newtonians call science or believe in nature and nature implementing science because this is what nature uses instead of Newton. These 4 phenomena used by nature in nature disprove Newton completely. Science never shows how these 4 works because they can't. I prove that the Universe consists of two cosmic substances which is material and space albeit either in a denser cosmic liquid such as light and cosmic clouds or less dense cosmic gas such as dark outer space Gravity is the movement of space within space. The entire Universe is relevancies formed by differentiation of density. Movement brings comparable density differences. This concept destroys the myth of mass pulling mass.

A Black Hole is much denser than the sun because it spins faster than the speed of light and the inside of a galaxy is denser than the outside since the inside spins faster. Everything in the Universe moves. By turning material contracts space thus reducing space surrounding objects and this leads to increases seen as space expanding. The faster a star spins the denser is the star and therefore the denser a star is by fast movement the smaller the star becomes in overall size and space. Since the area called outer space does not move the density reduce as stars collect the density by contraction of space. As outer space does not move it loses density in relevancy to material growing in density so it seems as if space expands in relation to material seemingly growing. This is relevancies changing as it seems the earth is getting larger at the circumference and it seems the moon is growing further apart from the earth but it merely seems that way because it is density moving by gravity from space unoccupied to space occupied. But space can't grow because the Universe holds all the space there ever can be and so expanding space is impossible because whereto will it expand; space has no where to go!

I wish to make one fact very clear...the way that Newtonian science shows the Universe work is not working anywhere. There is weight but not mass. Weight weighs and according to science mass pulls. I base my work on formulating the working process of four cosmic principles in Nature. These are:

The Coanda effect, which is the way, the atmosphere forms or liquids respond to solids moving.
The Titius Bode law is how planets use a very specific ratio to arrange their allocated positions.
The Roche limit is amongst many also the law that applies to what we call as the "sound barrier".
The Lagrangian points are why the different layers that form the atmosphere around the earth.

I did not discover these phenomena because science knows about these phenomena for a very long time and in some cases even for hundreds of years. Science knows they apply and where they apply. When science discovered or allocated missing planets they used the law applying such as the Titius Bode law from which they deducted positions that they knew in that circle according to the planetary layout that the law predicts there had to be a planet according to the Titius Bode law. Science did not apply Newton's formula to discover and locate planets but they applied these phenomena and especially applied the law of planetary allocation to discover the precise location the planets discovered after Galileo.

The response I get from a uniformed part of the public is astonishing. Believe me if there is one thing I appreciate then it is criticism backed by intellect. It seems the less they know the more they wish to teach me on matters about how they were taught how Newton presented cosmology. I am not interested how you think Newtonian explanations work in the Universe because Newton never applies in any form in the Universe. That which you believe explains Newton; it explains your personal brainwashing and how you convinced your mind that Newtonian impossibilities are possible. I am not interested in how you see Newton works in nature because nature has no room for any Newtonian principles. If you think nature uses Newton you better start reading up on what truly goes on in nature. Those that criticize me usually start off by saying "I did not read the entire book" and then goes on to try and teach Newton to me. I know all about Newton because I have studies Newton since 1970. The reason why you did not read the entire book is because the truth about physics goes beyond your mind limit. Don't think I insult you, every Newtonian out there that has a PhD in physics have the same problem; nature is above their understanding and the Newtonian simplicity calling on to enforce magic powers they can manage and therefore they go for Newton…not because Newton is correct but that is all those physicians can understands about physics. Newton saw forces and he was an alchemist by trade and therefore he had to believe in magical forces and powers no one can explain such as gravity. Those Brainy Newtonian Masterminds are fortunate because they can repeat after Newton albeit incorrect or disastrously wrong and that is because they **understand Newton.** I present you with facts from nature and not Newton!

Every one in science knows these phenomena are there and are in place and they rule the orbit set-up by forming the planets' gravity. The solar system functions according to them. These four laws on planetary motion that is used by nature at this moment and has been in place since time began are what apply and they dismiss Newton. If you argue with me about Newton being correct you better take your case to God or the solar system because the four cosmic phenomena is working in nature and nothing Newton said is applying in nature. This is a truth and a fact and a foregone conclusion and can never to be in doubt. Get educated in cosmology before you try to get wise and become opinionated about Newtonian correctness.

Brainwashed as you may be in believing Newton you can't either side with my view or decide on Newton because it is not a case of choosing between Newton and me. I'm out of the picture! It is either telling nature to listen to Newton and change what is in place or read and see what is in place and what is applying in nature all along. The phenomena are what we find to be used in the cosmos while Newton is in the imagination part of the minds of scientists and nowhere else. If you don't believe me and if you wish to discredit me first find out a little more about science.

Then deflate your ego as to what you think you know. Go and study what is in the cosmos and how the cosmos works before giving me a lecture how Newton works. I am not interested in how you were brainwashed and how you now are programmed to believe in the impossible. Go and find out why the rings are circling around the large planets such as they do around Saturn and why the material is formed as rings. If it was "mass" of the sun pulling the "mass" of the comets and therefore the comets are drawn to the sun then go find out why comets don't collide into the sun but circle around the sun and disappear into the darkness of space. Go find out why the moon is departing from the earth notwithstanding that according to Newton the moon and earth must come closer.

I found out why the four phenomena are in place instead of Newton ideas. Science never mentions these phenomena because science can't use Newton and explain these phenomena or use these phenomena to prove Newton. These four phenomena that the cosmos uses as we speak have been in place ever since the Universe formed. Since science can't explain the phenomena and the phenomena destroy the credibility of Newton science avoids these phenomena as if it brought the plague. You can't choose between Newton and me because I did not put the cosmic phenomena in place. All I did was doing a study since 1977 to formulate how and why these phenomena work and how these phenomena keep the cosmos and the solar system working. I am the first person in human history able to show why they work.

We are all been brainwashed for centuries to believe Newton. Should your brainwashing kick in and you have an axe to grind with me about what I say, then first prove these phenomena are not in the cosmos and are not applying to form the laws that the cosmos put in place as gravity. I only found out how they work and why they work and I did not make the phenomena work. All those clever stooges that have so much to say even before they read my work first learn what is in place before getting opinionated. It is not me that takes Newton out of the cosmos and replace Newton because Newton was never in the cosmos to begin with. While the four phenomena are in the cosmos Newton is in the imagination of scientists and in the mind control they placed on students? When you argue with me it only shows how little you know.

I am not fighting science or the credibility of Newton or what might be true or not true but I am fighting centuries of brainwashing unleashed on everyone young and old and I have to dismiss the mind manipulation, the brainwashing and the systematic mind control that those teaching science inflicted on us all. My fight is not about what is true or not true but what is accepted as culture and which was not even once been proven by science. Please do read on to investigate before you have the enormous wealth of wisdom that fires you up and you become self-opinionated. If your brainwashing starts to control your common sense please don't share it with me but go and put Newton in the cosmos and show where you found Newton was. I am not interested in the method by which they brainwashed you. First show me where it is in the cosmos that Newton hides and where it is that Newton's ideas can be located.

Before the French revolution people were force fed information when everything was told to everyone and it was expected of the faceless mobs and working class forming the illiterates to believe what they were told by the educated that told everyone what to believe and what to think. So since then what have changed because after everything the people still can't think using their own minds! When I came along and was not impressed by others telling me and did not take fore granted what others told me what to believe I became part of what was the intellectually deprived because **I did not understand Newton.**

When the urge overwhelms you to confront me about the honourable scientists all working in physics and how I am suppose to kneel before their supreme and ultimate wisdom allow any of them to prove mathematically how the four cosmic phenomena work and why do they work.
The phenomena are:
The Coanda effect, which is the way, the atmosphere forms or liquids respond to solids moving.
The Titius Bode law is how planets use a very specific ratio to arrange their allocated positions.
The Roche limit is amongst many also the law that applies to what we call as the "sound barrier".
The Lagrangian points are why there are different atmospheric layers forming that are around the earth.

I even have had astrophysicists hitting me on the knuckles while holding a Bachelor in physics or some Bachelor of Science degree proving they never heard of the 4 phenomena. Criticizing me publicly proves that they know Newton while they know nature not. Nature, not me because it is nature that uses these phenomena not to disprove Newton but to make the entirety you call the Universe function as it does.
If you ask me what is the difference between how I see gravity work as being the principle which I prove by going to nature and how Newton's gravity works, which unbelievable as it is but I also prove Newton has never been proved by science to form part of nature. What makes the difference between reality and science is Nature does not use Newton in any way or form. Newton says objects pull by an unexplainable magical force called gravity while I say it is round objects that rotate. As it turns it forms Π and by collapsing Π the space around the star / planet that space forming an atmosphere compresses and thereby collapses from $21.991 / 7$ to $3.1416 / 1$. As it turns it divides 7 from $21.991 / 7$ by 7 to form $7/7 = 1$. As the direction rotate Π changes the travel by $7°$ and that $7°$ is $7 / 7 = 1$ or $\Pi = 3.1416 / 1$. Then by getting reduced it compress everything in that space as the entire space the object holds and claims condenses. Space reduces and Newton said objects in space pull each other. While I prove nature, Newtonian science cheats, corrupts and manipulate nature to make science work in ways nature doesn't work. There is a link between space that compresses and material that spins and condenses space.

That statement I prove and I prove undoubtedly by example. There is no pulling of any object on any other object as Newton said. Gravity does not work like a magnetic field but works in principle by the object rotating... But it is space that compresses as rotation reduces in a principle called the Coanda effect. It works by material objects producing movement in rotating that condense space around the object of to liquid. Gravity turns space surrounding the rotating material from gas to freeze it into liquid. Outer space is a gas that turns to atmosphere that is a liquid. This compressing of space is the Coanda effect. The difference between what my approach must be when using nature and Newtonian's approach that is unsupported by nature is one Universe away from each other. I show a functional Universe and Newton show what they call the mysteries of science in the Universe. I prove everything that Newtonian science this far could never prove. I prove the Universe applies four keys by which gravity works instead of unexplainable magical forces that pull each other and this no person (not even Einstein) this far could ever explain. There are 4 principles applying gravity and forms a value of Π as a circle rotates.

By implementing the 4 cosmic principles I show in the book thereby I show how nature works with the 4 keys while I show how science falsifies facts to make science seems to work by magical forces pulling. What I show was never printed before by any person and is understood for the first time in human memory. These principles redistribute cosmic density from liquids to solids. Gravity is space moving as Π. If you read and as you read you have the opinion that you share the following view about science then you are brainwashed past the point of having your personal mind. You have become what they tell you to believe. Here is some remark those geniuses come up with hen they criticize me without knowing their arse from what I say. This is some of the comment indicating the stupidly I endure. A reader by the name of Leppad says he DISLIKES this book. This is why he dislikes this book and this is his informed opinion!

This author is wrong and has no understanding of science and he does not even know what the scientific method is. I will tell you why as briefly as possible. Not only do I think he is wrong, but I think he is doing a disservice to people by deliberately misleading people and muddying the waters thus causing great harm to anyone looking for truth. Apart from the irrelevant diatribe that has no bearing on the subject, the author's main premise seems to be one of conspiracy by the scientific community to deceive the public. His main proof is his lack of understanding of basic scientific ideas. He states that the equation F = [(m1 * m2) / r^2}] * G clearly shows that the larger the mass, the larger the force is and the faster the mass should fall relative to a smaller mass. This seems to be true if you have a myopic view of the equation but it is wrong for precisely this reason: Let's simplify the equation so you can check this in your head. G is just the gravitational constant therefore can be completely ignored since no matter what the numbers are before it, it will not change the relative results at all. So let's just ignore the G, or call it 1 if you will. The radius (r) squared can be assumed to be the number 1 for the purposes of comparison and can also be completely ignored since anything divided by 1 = the thing itself. This simplifies the equation to F = mass1 * mass2. Clearly, this equation shows that the larger the mass, the greater the force. However, Force = mass * acceleration. Solving for acceleration, the equation becomes acceleration = force divided by mass. Using this equation you can see that no matter how large the mass and hence the force, the acceleration is exactly the same for any object no matter what their relative masses. So the author is completely wrong when he wants to claim that the equation F = (m1m2/r^2) *G proves that a truck falls faster than a ballerina falls faster than a frog. If the acceleration is the same, the velocity, or rate of fall is exactly the same for all objects. Contrary to what the author claims, the truck, the dancer, and the frog will all hit the ground at the same time (in a vacuum). I did not read the whole book because it rambles off topic and often goes on irrelevant tangents making it extremely difficult to find a common thread or theme of any kind, but I plan to finish it just to see how it goes. This author needs to get an education, exercise critical thinking skills on his own stuff, and not think he is smarter than the entire scientific community.

This is the epitome of the power of brainwashing by submission to mind control. I am not interested in the way they arranged your mindset or how they made you a puppet of the way they think you should think. I am not interested to know how small your concept is on science because it is stooges such as you that keep those brainwashing you in the position where they can put their mind control on you. Use Newton's mass and explain why the comet does not hit the sun if the mass of the comet and the mass of the sun are responsible for pulling the comet to the sun. Then with all your intellect you show above, please use the formula you describe so passionately to explain why the comet then pass the sun and speed of into the dark unknown. If the mass pulled the comet, then what mass pushes the comet back into the abyss? What makes the comet go into a cycle and return after a specific lapse of time? What drives the comet away from the sun if the mass pulled the comet towards the sun. If idiots like you only will not think so much of your personal brilliance as to show everyone how you were brainwashed into accepting anything as long as those brainwashing you are fooling you into believing that you are clever. Moreover explain how mass lines stars up in this following example of planetary layout and then give me a lesson how Jupiter as the biggest with the most mass finds itself in the middle of the solar system?

Are you able to see the big planets are in the middle and the small planets are the closest. Have you ever heard of called ==The Coanda effect==
==The Titius Bode law==
==The Roche limit==
==The Lagrangian points== . If you have you would not talk the hogwash you have been explaining and if you never heard of it get off your arse and stop pretending to be clever and get informed before you get so wisely opinionated.
You and everyone else can contact me at mailto:info@questionablescience.net or at mailto:orders@singularityrelavancy.com If you feel you don't agree don't waste your time with this then go to a book that holds this theme at a higher level.

I WISH TO DEFINE THE CATOGORISING I USE AS PART OF THE BOOK.
I have the utmost admiration for Scientists and I shall never dream of placing me in the same category as academics mainly because of their intellect and achievements. They pushed their corrupt conspiracy of a hoax they present as science and which they further by brutal brainwashing through 300 years of never getting detected and that in itself is an achievement unheard of in human history. That achievement is most brilliant and no religion of magical mysteries in the past could ever match the Newtonians. Every time I go against Mainstream Science which is another name for upholding Newtonian blindness I am told I do not seem to have the intellect or mental capacity to **_"understand Newton's classical mechanics"_** and then because of my limited vision on physics I should know my place and retire to a dark corner where I would then silently and quietly vanish from earthly records. They forever tell me there are two positions on earth: those with the mental capacity **_to understand Newton_** and then there are those in my sector **_that is mindless to the point of not understanding Newton._** In that sense there are two classes, the clever ones that **_understand Newton_** and then me, the mindless that just cant **_understand Newton._**

Can you think of any religion that spread its entire belief on self-promotion without ever proving one validation? Can you think of some gospel preacher going around telling facts as if submitting the utter proof and the entire base is not even valid as a semi believable rumour? …And when there are Doubting Thomas's you turn the untruth of the system to reflect on their stupidity? They are brilliant! They are the true masters of deception and then they think Devil worshiping is deception. I am going to put a statement and to prove how brainwashed all person are, I think not one will catch on to what I am saying. All things fall equal notwithstanding size. I giant battle tank will fall at the same pace as an empty drum falling next to it. We see this on television almost everyday in all sorts of advertising. Therefore when things fall weight and size does not carry any differentiation as to the falling process. Yet it is **_"mass"_** that pulls the falling object to create the gravity or the fall. Everything falls the same in free fall. Then you explain to me how the F$&@ck can **_"mass"_** have anything to do with what causes the falling of objects. The fall is neutral in size and in weight but linear movement brings about the stopping of the fall and launch flying? Those physicists don't prove anything because they brainwash everything into submitting.

To substantiate this segregation I use some referring to place distinction between the highly schooled super trained academics that spent most if not all of their lives in preparing to further their minds, filling it with the same void they fill the Universe and calling it "nothing". When I asked where is more nothing: Between Pluto and the sun because Pluto is the furthers from the sun holding the most "nothing" between it and the sun or in the centre of the sun because there is nothing standing between the sun's centre and the centre of the sun, I was discredited as incoherent and irrational. I tip the opposite of the scale as I spent little time repeating the brainwashing they subdue every student with to believe in the norms taught as the official policy in learning and education I have to be on the "other end". I don't believe their crap and tell it as I see it and therefore I am dumber than a pig, or that is their opinion. From where I stand and admire those in science, I can only see intellect as they fooled every person on earth for centuries non-stop: and moreover that achievement is presented as the academic's common denominator. If that is the common denominator used on the one side, fooling everyone by using unsubstantiated rumours and gossip and putting that as the joining factor, then on the other side, which has to be *"my side"* must then be the class of stupidity. To those forming the brilliance in science and their class such a remark would be an insult but to me (and therefore my class) it rings truth and that makes it not an insult but a norm we should except and learn to live with. I would rather be stupid and not **_understand Newton_** than be **Brainy** and believe I **_understand Newton..._** how stupid must I be before I would be able to **_understand Newton._**

It is rather a pity that while the SUPER CLASS will never say it to our faces; the SUPER CLASS is strongly of such opinion that we on the other side of the Universe have no minds to think in any way, and it is therefore our duty as much as it is our absolute privilege to except what the SUPER-EDUCATED, the ones occupying the informed side of the Universe inform us to what we should accept and the SUPER-EDUCATED live by that idea. As I said I have to live with it too and if I am the ill literate, then the SUPER CLASS must be the SUPER-EDUCATED; where I am the class amounting to stupidity the SUPER CLASS must be the Brainy Bunch. It all comes from the fact that there is such a huge differentiation between us. Those that **_understand Newton_** is therefore Superior and I, that don't **_understand Newton_** are of the lesser blessed. To distinctly point to grouping or class or whatever the readers wish to consider the division there are between the SUPER-EDUCATED and me I refer to the SUPER-EDUCATED side of the Universe by the names I use above. Further more when I refer to mistakes that I do prove to be mistakes in the book as we go along I refer to it as Xepted mistakes to clear another distinction of

necessity. In short I don't **_understand Newton_** and therefore I am stupid and they **_understand Newton_** and therefore they are brilliant and what I present must hold the categories in such class divisions.
I am more than open for a debate but get sensible and stop being brainwashed! First explain what drives the comet away from the sun if mass pulls the comet closer. If you feel giving me a lesson in Newton's correctness I suggest you first go for a big lesson in cosmology because I am not taking Newton out of the cosmos, he was never there in the cosmos but I am taking Newton out of your imagination and I am resetting your brainwashing that was so brutally inflicted on you. The gravity the Universe holds is the distributing of space or movement where with all the rotation forming the visible gravity it also brings along a certain amount of linear gravity to maintain the circular gravity position it wants to hold in the cosmic balance. The duel movement of gravity forming the allocated relevant position provides ratio of singularity Π accompanied by the same value but in movement by the square thereof Π^2 and repositioning the structure as a star.

I now explain gravity again and relate this explanation to the sound barrier functioning.

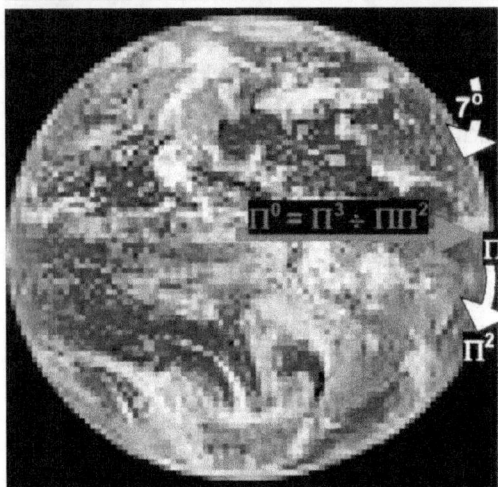

Gravity forms by association or relevancies that holds a value of coming towards the centre from the sky which then is 7(which is the curve of the earth) (Π associating with the curve of the earth) and Π^2 (associating with the moving of the earth) This is part of the Titius Bode law forming the measure of gravity which is not by mass but by forming Π.

$7\Pi\Pi^2 \times \Pi^0$ to $4\Pi^0$

Then when a line is drawn from the centre of the earth another value comes in place referring to the centre of the earth Π^0 and the curve of the earth Π. The value of Π is then shared by both disciplines (association from the sky to the centre ($7\Pi\Pi^2$) and association from the centre to the sky ($\Pi\Pi^0$). In reality it is ($7\Pi\Pi^2$) but I am not explaining this.

This is what there is and that is all there is. The measure of mass forming gravity clearly plays no role in allocating the positions of planets as Newton declared it must do. The entire idea that gravity is a magical force created by the value of mass is as unbelievable as the dogma is of those presenting this idea. Please use what the solar system provides to confirm what Newton says is in place when he says mass forms gravity. Science would rather accept Newton where there is no proof of Newton ever being correct than to admit Newton's incorrectness.

In any circle the radius divides 3.1412 times into the circumference. We also know that when the radius holds a value of 7 the circumference holds a value of 21.991. It is not 22 as some mathematicians try to minimize the accuracy but it is 21.991. That means the ratio is one circle will allow 21.991 dots to form a circle in having the circle move in ration to 7 point forming a straight line. That is the significance of time.
Every layer is the depreciation of space by 7 going singular or 1 as Π revaluates from with

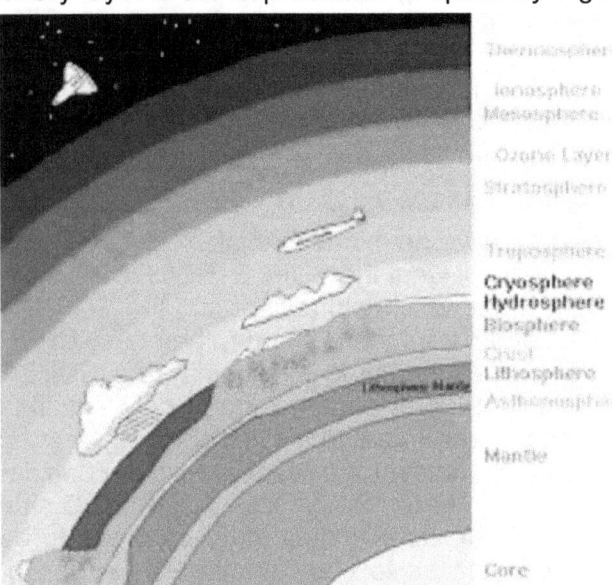

space $\Pi = \dfrac{21.991}{7}$ to space went singular. $\Pi = \dfrac{3.1416}{1}$

That reduction indicates that gravity is the spin that revalue Π from $\Pi = \dfrac{21.991}{7}$ to $\Pi = \dfrac{3.1416}{1}$. Every point below the point above is representing singularity where the top one above the one below is Π. From the centre a line forms holding singularity Π^0 in relation to Π and notwithstanding the distance of the line the point below represents singularity Π^0 to the point above that is Π. Any line will start with Π^0 and this will continue holding Π^0 until it concludes with a circle forming Π. Therefore a line runs from the earth centre as $\Pi^0\Pi^0\Pi^0\Pi^0\Pi^0\Pi^0\Pi^0\Pi^0\Pi^0$ until it reaches Π where it then allocates mass as a factor holding the mass in relevancy to the earth being 1 or Π^0. All aspects of movement holds a circle forming gravity and that is $\Pi^0\Pi$. Gravity is the movement of Π as Π

going to Π or $\Pi \times \Pi = \Pi^2$. There is no square in the Universe because the Universe is square by motion.

A SCIENCE COVER-UP

A SCIENCE COVER-UP is a range of four books that some will be published by using e-book format and later in paper print. We don't all deal with facts in the same manner. These books are the same book where some is more comprehensive developed offering better-detailed facts and others are less bewildering bringing more down to earth facts.

I have four versions of precisely the same book. It deals with science and science has a wide-ranging field of interest. The one book is shorter or longer which means it is better or lesser developed than the others. Some people are easily convinced and need little information to reach a conclusion whichever way. More information would not change their opinion either way. I did this categorization with those close to me in mind because we as a family are as ordinary as everybody in the world. Then others require more information to be informed on any subject as to be convinced but don't like to be overburdened by much explaining and in depth study results. Whatever you choose is your choice but make sure that you can deal with the information that that specific book presents you with and that you can divulge and understand what the information reveals. The message is comprehensive and the arguments are all new to you.

The media in every form is overflowing with all sorts of corruption going on in commercial banking, in politics, in corporate pharmaceutical medicine, in hospitals and every other industry...Then what about science? Why is science that Lilly white they make it to be or is it? This book is the whistle blower on fraudulent practising going on in science. This book informs about shocking detail everyone misses and reveals how deep the misconceptions are while this book exposes details of blatant mind bending brainwashing practises science are committing to unsuspecting students. In this book there is a business offer that can be very lucrative in participating to uncover dark secrets that large money hides to keep the world slumbering in fogginess and twilight. Reading this book will inform any reader on any level more than any other book did before and reveal a financial offer that will not repeat again. You can become part of the battle against science fraud and be financially rewarded.

Informing about Corrupt Science.

ISBN

http://www.singularityrelevancy.com/

You are About to Uncover

A Science Conspiracy

© KOSMOLOGIESE EN ASTRONOMIESE TEGNIKA

www.questionablescience.net

WRITTEN BY PETRUS S. J. SCHUTTE (Peet Schutte)
©KOSMOLOGIESE EN ASTRONOMIESE TEGNIKA

All rights are reserved.
No part, parts or the entirety of this book may be reproduced by publishing, electronically copied, duplicated by whatever means that form reproduction or duplication, without the prior written consent of the copy rite owner.

I am Petrus Stephanus Jacobus Schutte going by the nickname of Peet and I am a married male, with a sane mind and I hold very sober habits being a lifelong teetotaller therefore my mind is clear. I am as sober as a judge but with a difference and it's that I am sober while judges are not. By nature I am frustrated beyond what any mind can take. I am as mild mannered and friendly as any rhinoceros with acute molar tooth infection and I am as gentile as a lion that got his tail pinched in a vice. Now that I have introduced my friendly and mellow side whereby I try to expose my soft and tender nature and soft underbelly let's get to my work to show my crocodile teeth.

If you criticize me in favour of Newtonian science what that proves is that you are brainwashed about science and you are not educated in science. You don't find Newton in nature because nature never used Newton or any aspect of what Newton said nurture is. If you go bashing me it proves you are totally uninformed about what nature uses and you never progressed beyond being brainwashed. Arguing shows you blatant stupidity about the reality of what is truly present versus what you were told is present in nature. I suggest you read and for the first time get educated and not suffer mind control by academics brainwashing you in believing in magical forces that never existed before or presently. Do as I did log ago; act intelligent and study what nature uses and not what Newton said. You're going to go in the correct direction following nature. What you are about to read is the mother of all the conspiracies in science.

Every conspiracy that Science ever thought up such as the Critical Density Conspiracy or the Dark Energy Conspiracy, or any conspiracy connected to science is in place to protect this theory they hide from anyone outside physics from becoming known. All of the conspiracies in place in the past or at present do the hiding of the Mother Conspiracy so well that in three hundred years no one that is not part of science got a sniff about what the Mother Conspiracy entails. The Mother Conspiracy is in place so that students in physics are brainwashed by instigating the deliberate sanctioning of Mind Control through their practising of enforcing Thought control they unleash on students. I reveal the Mother Conspiracy and I show why it is in place. I prove that the Mother Conspiracy is in place. I show how every student including you reading this has been brainwashed to believe science is true and to believe in science. This book is for free; I make no money so why would I be dishonest? When you download it you are going to get the nastiest surprise of your life. Then you can also read how the cosmos truly apply science and read Explanations of what was never Explained before and how the explained factors interlink

What you are about to read holds nothing new and definitely nothing about my work. I introduce what is in nature and I show what nature uses and therefore when you disagree with me you don't disagree with me because you show your personal inadequate level of truthful information about what you should know if science were not fooling around by brainwashing you into believing a fairy tale story about Newton while they know you know nothing. As I said I say again, this is what nature or God Almighty uses to build the Universe and when you eventually read my work you will find out why these building blocks work the way it does, how does it come about to function as it does and the entire process how the Universe came about from the first original dot. By the way that is why you can see the entire Universe with the space your eye holds and that is why you are within the centre of the Universe as all the light comes to you where you are. It is because I only show what you were suppose to know if science didn't cheat you witless that I offer this book at such a low price. Should you wish to purchase my work to find out the mysteries driving the Universe and what is the Universe you will have to pay more. I'll show you why you can see what you see I studied physics and in my first year of studies I came upon a mistake concerning physics. The mistake is so deep rooted and to get to it one must go back to ancient times.

I have progressed far beyond what is needed to get on friendly terms with persons judging their position as important people because others think they are important people. I got use to the idea that who ever is in power of whatever is corrupt to the core and no less are the department of Physics, Astrophysics and philosophical physics. I saw how absolute power absolutely corrupts They are as much gangsters a politicians and generals and Clergy of all dominations are. The more you give them respect the more they act as if they became God Almighty that rules the earth. When you read what you are about to read it is not the hallucinations of a drunkards mind. I introduce myself because I know you have never heard of me and there are hundreds of physics academics that keep it that way. It is because while they hide cover-ups in physics I reveal it.

If I tell you science is built on fictional fraud you will put this down and believe I am as loony as a hatter. If you think I am as mad as a hatter you are in for a brutal surprise because science has as much criminal activity to cover than does loan shark bankers have however science is respected. When you have read what I uncover you will agree with me in the way I see myself as very responsible being down to earth and am not well liked because of my straightforward personality where I say what needs to be said to maintain honesty above friendship and what I say is always not infusive or congenial to make friends. As should be evident by now: I wrap no feelings in cotton wool for the sake of peace. I can't sue for peace while I battle the corruption of science.

Science hides an entire Universe behind "nothing". My roommate told me it was Newton that said what I disputed and I remember telling him I don't care who this Newton fellow is but he is wrong and ever since then I stuck to what I said. That discovery made me disagree with the establishment forming principles in physics. Later in time I have detected much more than a mistake. When you read my work and banish your cultural brainwashing you can judge my work. You then can see what a huge Pandora's box I have discovered. Science is believed because of mind manipulation and brainwashing going on forcing students to believe notwithstanding truth.

I try hard to be honest even though I am poor. I am poor because I present the truth. If I did try to accommodate Newton and the falseness Newton embraces I would be embraced by science but I just can't underwrite Newtonian incorrectness for the sake of money. The proof of my honesty is that I am one of the poorest people in society. If I want a cold drink I have to ask my wife because I have no income trying to introduce the truth to science. I have to fight an ideology of corruptness.

Yet for all the good the exposing on my side did me I still write books that don't sell because I try to convince people about mistakes no one on earth are aware of and therefore no body bloody cares except me. I am the only one that can see a mistake and seeing it in the full consequence thereof by what it holds to the entire human race it is frustrating me senseless. I see the mistakes because I see how to correct the misleading dogma in science and by correcting the mistakes I show a clearer science. He total size of the conspiracy to corrupt goes beyond what any mind can cope with at first. At first you will think I am exaggerating to try and make a profit but then look at the price of the book I sell and try to find any profit in a book selling at such a price. My effort is showing how to correct the mistakes I show they hide.

I have spent a lifetime researching how nature works and I discovered how the building blocks of the Universe works. I did not discover the building blocks but I di discover how the principles work, how they work and what results from the working of the principles. While Newtonian science suppresses the importance of the phenomena in favour of falsifying the correctness of Newton I am blocked revealing what I discovered. As long as they legitimise by conspiracy what they claim is absolutely godly truth about Newton's ideas and trying to maintain Newton as being supported by nature I am chained to frustration by a falseness science presents. They hold science out as if nature completely supports Newton and nature never supported Newton in even one idea the man had. I can show how the Universe works but as long as Newtonian science legalises Newton's falsification of facts this conspiracy to uphold Newton. When I challenge those in office I get blown away because they argue Newton needs no proof since time proved Newton. They don't have to prove Newton while nobody ever proved Newton as long as science exists.

With the hardship and poverty my family went through especially my children while growing up I can't hate anything more than those in astrophysics because they all should be in jail for uncompromising dishonesty and mind managing corruption and then they are on top of the list of those presenting honesty. They block me with fraud and corruption. With their covering of fraud they put me through hard times and poverty while they ride on the wave of complete holiness.

According to my approach I mention these facts to establish beforehand that I am not a danger to society. Although I am the only one in the entire world that totally disagrees with modern mainstream science and mostly on the thinking of physics yet, I am not criminally insane. I would not attack your dog before your dog will bark at me because I hate your dog. I would rather attack you and not your dog after he barks at me just because I most probably won't like you but I do like dogs. I say this to convince you that I have never been jailed in all my life on grounds that I attempted to mislead or fool or tried to mislead anyone, as you can see. I try hard to be honest even though I am poor. The proof of my honesty is that I am one of the poorest people in society.

I am so poor because I write books that don't sell because I try to convince people about mistakes no one on earth are aware of and therefore nobody bloody cares except me. I try to introduce honesty in the midst of fiction ruling science. I bring the truth and remain poor while Dan Brown introduced fiction as truth and a lie got him rich. I am the only one that can see mistakes and seeing it in the full consequence thereof by what it holds to the entire human race it is frustrating me senseless. I see the mistakes because I see how to correct the misleading dogma in science and by correcting the mistakes I show a clearer science. Take it from me that writing about science while trying to convince people about mistakes everybody but me believes is correct and therefore not selling books is not a very profitable enterprise and is frustrating at any level. If I was less honest and went about bullshitting everyone about the accuracy that Newtonian science portrays in the modern era it would at least sell some books because some dinosaur Newtonian physicist would be pleased about it but then I was a cheat although I would have been richer but being as poor as I am in favour of trying to be as honest, in that I am can't be a cheat. I studied physics and in my first year of studies I came upon a mistake concerning physics.

I am held back by powerful people because Science hides an entire Universe behind "nothing". Years back my roommate told me it was Newton that said something that I disputed and I remember I told him I don't care who Newton is but Newton is wrong and ever since then I stuck to what I said. That discovery made me disagree with the establishment forming principles in physics. Later in time I have detected much more than a mistake. When you read my work and you banish your cultural brainwashing then judge my work. You then will see the huge Pandora's box I have discovered. Science is believed because of mind manipulation and brainwashing going on forcing students to believe notwithstanding truth.

It is obviously clear why I should be the one that can gain no grounds or benefit from finding at least one case where there is one academic somewhere showing one bit of doubt about Newton and therefore are prepared just to read my books from start to finish without their commitment having Newtonian bias that is interfering with their judgement. Not once could I find one academic that would sit with one of my books from start to finish and read past the page where I start to show Newton's defects, although I am so obviously correct on every matter that I state. I am washed off the Earth for I show little regard and even less respect for the consecrated sacrosanct hierarchy of Newtonian wisdom and for that attitude I find no ear in the world of science prepared to listen to my views which they deem as clear insanity.

At first I thought every physicist was a champion in fight for the truth. Later on as my personal naivety about their honesty diminished I could have some blame about my conduct as I addressed their mighty Holiness as their Royal Highness The academic professors in astrophysics and in physics found any misunderstandings they could use against me about the misgivings I had about Newtonian science and they might now condemn me for showing a lot of antagonism and disrespect, showing that I had an attitude, but that was not the case at the very beginning. I am sure that it could at this point influence them where I don't kneel as low and shiver with fear when they enter the building with their wisdom surrounding them like a whirlwind, but this was not the case a few years back and certainly not at the start. In the beginning I showed much respect and that got me nowhere very quickly but also painstakingly slowly and with it the overall experience I went through was expensive. I started attacking their religiosity with venom. Now I go into detail and prove what fools those honourable Academics in physics are and how they corrupt the young mind to brainwash the young and vulnerable in accepting the detestable criminality of lies and deception that they call astrophysics. I challenge any one to show the correctness of Newton's view about gravity in the face of the evidence I am about to bring. I charge students to challenge the academics with the evidence that the academics present to portray what they advocate as being religiously correct. In order to reach the heart of gravity one must discover the heart of gravity. I challenge those academics that are pretending to be so sure about Newton to prove Newton!

At first I was in awe but as those Brainy Bunch could never prove Newton I started attacking their religiosity with venom. Now I go into detail and prove what fools those honourable Academics in physics are and how they corrupt the young mind to brainwash the young and vulnerable in accepting the detestable criminality of lies and deception that they call astrophysics. I challenge any one to show the correctness of Newton's view about gravity in the face of the evidence I am about to bring. I charge students to challenge the academics with the evidence that the academics present to portray what they advocate as being religiously correct. In order to reach the heart of gravity one must discover the heart of gravity. Those academics that are pretending to be so sure about Newton I give the following challenge:

Hidden under a cover of "understanding Newton" or "not being able to understand Newton" tutors in physics force certain incompatible arguments to join that which never can join and while joining also make sense at the same time. In this letter I challenge those presenting intellect and those that charge the highest form of respect in our communities and those in charge of the most dynamic part of society and those who stand beyond and above any form of suspicion to earn our respect for once. I charge those that personify truth and are the very same persons that I accuse of betraying the ones trusting them.

At first I thought every physicist was a champion in fight for the truth. Later on as my personal naivety about their honesty diminished I could have some blame about my conduct as I addressed their mighty Holiness as their Royal Highness. The academic professors in astrophysics and in physics found any misunderstandings they could use against me about the misgivings I had about Newtonian science and they might now condemn me for showing a lot of antagonism and disrespect, showing that I had an attitude, but that was not the case at the very beginning. I am sure that it could at this point influence them where I don't kneel as low and shiver with fear when they enter the building with their wisdom surrounding them like a whirlwind, but this was not the case a few years back and certainly not at the start. In the beginning I showed much respect and that got me nowhere very quickly but also painstakingly slowly and with it the overall experience I went through was expensive.

Again I challenge you to come forward and tell your students the truth about what I uncover in the articles that follows and as the articles progress by introducing information...then you explain to them how you deceived their blind trust in you as a tutor in physics. In physics the blame goes to students ability to "understanding Newton" or "not being able to understand Newton" but I show that in this case the blame should openly be dedicated to those that should be blamed, named and shamed and they have to defend their years of lying and contribution to cover the misconduct that was committed by them.

To find the truth you know is going on, then answer to your person in privacy and in all honesty the following questions that I put to you. It will gauge your state of brainwashing and show the amount of damage that you have suffered this far in your particular and specific case. Then go on reading and confront the truth, as you never had ever before. I wish to explain some part of my personality with the intention not to shock you with my deliberate brutal honesty no person ever enjoys. I once was a creature that showed respect but I grew to learn physicists deserve no respect! You will find I never compromise truth for friendship and in that light I say what they needed to hear and not what is wanted to please.

Meet the Newtonian physicist. IN THIS BOOK I try to introduce the reader to the brilliant Newtonian conspirator that has been dragging all of intelligent man by the nose for three centuries on a string. He promotes what can only be forces of witchcraft because he is unable to prove how the four forces apply the magic of gravitational pulling and yet he preaches the power that the forces generate.
The more the conspirator pretends to be an intellectual physicist the better a fool those conspirators become.

He looks sheepish because he acts sheepish because as he follows he never questions what he believes and brainwash students to do the same. Read how clever the physicists are in hiding their stupidity from students and the public alike. He puts everything down to "black magical "mass is pulling mass" because to him gravity is a force of magical proportions" and that he believes for he knows no better. He proves nothing to students but by enlisting thought control those teaching physics force students to believe in science by applying some cruel mind-bending manipulation and to have students believe science that has magic powers he makes them accept the unexplained. If you believe in their brilliant mathematical genius I advise you to brave your mind for a big surprise is looming… The truth is that entire cosmos formed and still rests because of four cosmic pillars and science doesn't even know this or are able to explain why and how these phenomena forms or are in place. These phenomena proves that there is no "mass" pulling anything because gravity.

I have asked this many times before but I ask again: Have you, the person reading this, ever thought how it is possible to see that much information that you see at night when looking at the sky with only your eyes? Ever thought about how you are able to see when you see everything in the night sky and how that much light information can fit into such a small space as your eye? Have you ever sat back and think what vastness in information it is that you see when you see the entirety of the Universe when looking at the Universe at night and what the size is of everything of that which you are able to see?

I write this where the reading is complicated so that it and the reader is forced to the reader to think. Many years outside my house one night with my stars through binoculars. I showed him with all my sons. Then something struck see and see it so clearly by holding a by putting a lens on space it magnifies with the naked eye a mere dot. I could glasses a flickering spot. How is this everything that I am able to see by just was making what I see more to see because I glasses and less light without glasses? That glass to magnify, but what do I magnify? We why would light magnify when using a lens? Moreover what science do I read into this. I taking Newton and Newtonian science same as you do when reading this. If there Newton then this night was the point. I did not dimmer. The entire vision I got was smile to think of the hours and days, months specifically takes an effort to read read slowly that would allow ago I was standing on my farm son Willem while I showing him the all sorts of stars and this I often did me...how is it possible to see what we lens. That means space is a lens because space. I could see a clear galactica that was clearly see binary stars that were with out possible? How can I see with my eyes standing and admiring the night sky? The glass saw less. Why would I see more of the light with makes what I see become more when using a call what we see light and that is too simple. But Light increase when using a magnifying lens? have to be honest that at that time I was still seriously and thought it had some value the was a point that got me to break free from see massive stars better and lesser stars emphasising Kepler's $a^3 = T^2k$. Sometimes I still on end when I tried to figure out how Newton got $a^3 = T^2$. I was still holding the idea that I had no mentality to grasp Newton's vision of $a^3 = T^2$. Then I realised what I saw was $a^3 = T^2k$. I saw the star a^3 in space $= T^2$ but I did not see mass because I saw distance k. I saw Kepler and nothing about Newton made sense. If the glasses reduced k then I could see space at the distance of k much better in the formula $k = a^3 / T^2$ and it was k that was responsible for making the Universe much clearer. That then made me have a look at Newton with an open mind and I did not try to locate mistakes at my door anymore but I put Newton under the magnifying glass. Then I realised physics started here at this point where I could see the Universe as big as it is with something as small as a nerve in my eye. The question was where does physics star? Does it start with some overgrown egocentric-maniac with a single minded mathematical mind only capable of processing data or does it start at this point where whatever is there reduces to a point so small the ration of reduction goes far beyond mathematical skills. Physics is about vision because the Universe in space is about vision, which makes space form by heat! That is science and not some outdated backwards theory made centuries ago

I write books that don't sell because I try to convince people about mistakes no one on earth are aware of and therefore no body bloody cares except me. I am the only one that can see a mistake and seeing it in the full consequence thereof by what it holds to the entire human race it is frustrating me senseless. I see the mistakes because I see how to correct the misleading dogma in science and by correcting the mistakes I show a clearer science. Take it from me that writing about science while trying to convince people about mistakes everybody but me believes is correct and therefore not selling books is not a very profitable enterprise and is frustrating at any level. If I was less honest and went about bullshitting everyone about the accuracy that Newtonian science portrays in the modern era it would at least sell some books because some dinosaur Newtonian physicist would be pleased about it but then I was a cheat although I would have been richer but being as poor as I am in favour of trying to be as honest, in that I am can't be a cheat.

If you feel I am exaggerating or that I am a mentally impaired asylum escapee that found a writing pad, ink and paper and that I now start to scrabble senseless suggestions to while away my social frustrations then I challenge you to prove that students are NOT being brainwashed to believe what they are taught in science. I started off being likable and being nice but the deeper I delved into the conspiracy the less I could care about who thought what about what I said because I am going for the truth as hard as I can. Their mannerism in blocking and frustrating my opinion when showing the mistakes in science convinced me about a Conspiracy in Science in Progress and this spurred me on to tell the entire world about their brainwashing students minds. By the manner they selectively withhold information when teaching science, amounts to deliberate brainwashing of students in physics by "normal" education practises. The new concept I wish to introduce puts all emphasis on space ands material is only space filled with material substance while other space is filed with non-material. In the end all space are equal but the movement it has makes the difference it presents in relevancy. All space structures hold in the centre most heat concentrated and from that centre holds all material owned by that structure. I can go on and on but heat in the centre couples gravity to space-time, just like Kepler said before he was spoken for on his behalf and without his permission or his agreeing to it. Studying Kepler helped to understand why the phenomena are there to begin with and that enabled to explain in some way…

I studied physics and in my first year of studies I came upon a mistake concerning physics. Science hides an entire Universe behind "nothing". My roommate told me it was Newton that said what I disputed and I remember telling him I don't care who this Newton fellow is but he is wrong and ever since then I stuck to what I said. That discovery made me disagree with the establishment forming principles in physics. Later in time I have detected much more than a mistake. When you read my work and banish your cultural brainwashing you can judge my work.

You then can see what a huge Pandora's box I have discovered. Science is believed because of mind manipulation and brainwashing going on forcing students to believe notwithstanding truth. It is obviously clear why I should be the one that can gain no grounds or benefit from finding at least one case where there is one academic somewhere showing one bit of doubt about Newton and therefore are prepared just to read my books from start to finish without their commitment having Newtonian bias that is interfering with their judgement. Not once could I find one academic that would sit with one of my books from start to finish and read past the page where I start to show Newton's defects, although I am so obviously correct on every matter that I state. I am washed off the Earth for I show little regard and even less respect for the consecrated sacrosanct hierarchy of Newtonian wisdom and for that attitude I find no ear in the world of science prepared to listen to my views which they deem as clear insanity.

At first I thought every physicist was a champion in fight for the truth. Later on as my personal naivety about their honesty diminished I could have some blame about my conduct as I addressed their mighty Holiness as their Royal Highness The academic professors in astrophysics and in physics found any misunderstandings they could use against me about the misgivings I had about Newtonian science and they might now condemn me for showing a lot of antagonism and disrespect, showing that I had an attitude, but that was not the case at the very beginning. I am sure that it could at this point influence them where I don't kneel as low and shiver with fear when they enter the building with their wisdom surrounding them like a whirlwind, but this was not the case a few years back and certainly not at the start. In the beginning I showed much respect and that got me nowhere very quickly but also painstakingly slowly and with it the overall experience I went through was expensive. I started attacking their religiosity with venom. Now I go into detail and prove what fools those honourable Academics in physics are and how they corrupt the young mind to brainwash the young and vulnerable in accepting the detestable criminality of lies and deception that they call astrophysics. I challenge any one to show the correctness of Newton's view about gravity in the face of the evidence I am about to bring. I charge students to challenge the academics with the evidence that the academics present to portray what they advocate as being religiously correct. In order to reach the heart of gravity one must discover the heart of gravity. I challenge those academics that are pretending to be so sure about Newton to prove Newton!

Why is the Universe depicted as a sphere...and why would that then be correct...
...how did everything become so much and so large...
...why did it start small...
...why does it grow from small to large...
...why was the start so small...
...why is it growing...
...where is it going while it is growing ...
...why is it any specific size...
...what was everything before that...
and why in creation would this lot then reduce again!!!

Gravity is

The Coanda Principle
That I explain

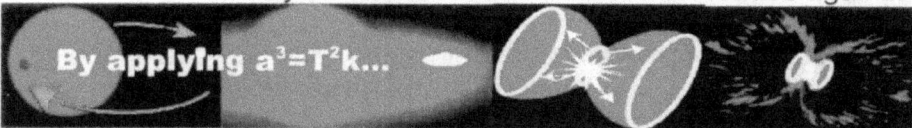

The Roche limit becomes self-explaining through the Coanda effect when using the Coanda principle in the practical

Every one in science throughout many centuries ignored Johannes Kepler because all saw him as some derogative of Newton...until now.
    Kepler introduced space –time but nobody took the time to acknowledge Kepler's introduction. Kepler introduced space a^3 – time T^2k and showed that it is space a^3 – time T^2k that is performing gravity by relevance of k. Are our centuries long ignoring of Kepler truly the answer...Kepler introduced gravity by principle but no one in four hundred years took any notice of the manner in which Kepler brought gravity into human conception and understanding?

Kepler calculated that it is the motion of space a^3 during the time T^2k that forms the gravity that is keeping the sun and all the individual planets apart but moreover gravity is keeping the planets in orbit. While every one was surprised but now accepts there is a growth in the Universe by which the Universe is expanding...for four centuries Kepler said that and no one took notice. According to Kepler the expanding is the normal trend that the cosmos will follow...that he said four hundred years ago...yet in spite of Kepler findings...science still clings to the idea that what keeps the Universe secure is contracting the force by the mass value that creates an attracting in the distance between the objects.
That is not what the finding of Johannes Kepler said.

It is not I that needs convincing that Newton is correct but it is the cosmos that needs the convincing since it is the cosmos that uses very different methods of applying gravity than those Newton suggested. If Newton is correct why does a comet not crash into the sun. why does the comet circle around the sun and move back into outer space just as it came. If it is mass that pulls the comet closer what then pushes the comet back into outer space. I can't be anti mass because there is no such a thing. So what is then pushing instead of pulling? Yet the comet goes away as fast as it came and the "mass" did not detour the comet in any way shape or form. The comet rushes into the black yonder as fast as it came form the black yonder and it will come back again because all comets rotate around the sun at precise cyclic intervals in accordance with following Π. Again I repeat if mass did the pulling of the comet what pushes the comet back into outer space because the comet is not crashing into the sun as the Newton formula suggests. Again, the cosmos ignores the importance of Newtonian views about "mass" pulling.

It is not me that needs to be convinced about using Newton but it is the cosmos because instead of Newton the cosmos uses four cosmic phenomena science is aware of but they clash with Newton and that is why they are pushed under that table in order to save Newton and Newtonian science.

At first I was in awe but as those Brainy Bunch could never prove Newton I started attacking their religiosity with venom. Now I go into detail and prove what fools those honourable Academics in physics are and how they corrupt the young mind to brainwash the young and vulnerable in accepting the detestable criminality of lies and deception that they call astrophysics. I challenge any one to show the correctness of Newton's view about gravity in the face of the evidence I am about to bring. I charge students to challenge the academics with the evidence that the academics present to portray what they advocate as being religiously correct. In order to reach the heart of gravity one must discover the heart of gravity. Those academics that are pretending to be so sure about Newton I give the following challenge:

Hidden under a cover of "understanding Newton" or "not being able to understand Newton" tutors in physics force certain incompatible arguments to join that which never can join and while joining also make sense at the same time. In this letter I challenge those presenting intellect and those that charge the highest form of respect in our communities and those in charge of the most dynamic part of society and those who stand beyond and above any form of suspicion to earn our respect for once. I charge those that personify truth and are the very same persons that I accuse of betraying the ones trusting them.

Again I challenge you to come forward and tell your students the truth about what I uncover in the articles that follows and as the articles progress by introducing information...then you explain to them how you deceived their blind trust in you as a tutor in physics. In physics the blame goes to students ability to "understanding Newton" or "not being able to understand Newton" but I show that in this case the blame should openly be dedicated to those that should be blamed, named and shamed and they have to defend their years of lying and contribution to cover the misconduct that was committed by them.

To find the truth you know is going on, then answer to your person in privacy and in all honesty the following questions that I put to you. It will gauge your state of brainwashing and show the amount of damage that you have suffered this far in your particular and specific case.

Then go on reading and confront the truth, as you never had ever before. I wish to explain some part of my personality with the intention not to shock you with my deliberate brutal honesty no person ever enjoys. You will find I never compromise truth for friendship and in that light I say what they needed to hear and not what is wanted to please. The above I explain by proving what happens when nature performs because I managed to formulate the working process of the four cosmic principles or building blocks in Nature. These are: 1) The Coanda effect
2) The Titius Bode law
3) The Roche limit
4) The Lagrangian points.
If you have never heard of these phenomena it is not that surprising because you either have to believe in Newtonian magic Newtonians call science or believe in nature and nature implementing science because this is what nature uses instead of Newton. These 4 phenomena used by nature in nature disprove Newton completely.

Gravity is the movement of space within space. The entire Universe is relevancies formed by differentiation of density. Movement brings comparable density differences. A Black Hole is much denser than the sun and the inside of a galaxy is much denser than the outside rings forming the galaxy. Everything in the Universe moves. By turning material contracts space thus reducing space surrounding objects and this leads to increases seen as space expanding. The faster a star spins the denser is the star and therefore the denser a star is by fast movement the smaller the star becomes in overall size and space. Since the area called outer space does not move the density reduce as stars collect the density by contraction of space. As outer space does not move it loses density in relevancy to material growing in density so it seems as if space expands in relation to material seemingly growing. This is relevancies changing as it seems the earth is getting larger at the circumference and it seems the moon is growing further apart from the earth but it merely seems that way because it is density moving by gravity from space unoccupied to space occupied. But space can't grow because the Universe holds all the space there ever can be and so expanding space is impossible because whereto will it expand; space has no where to go!

From the Heart of the Author, the following: Tyco Brahe, on his dying bed, kept on rambling to Johannes Kepler, begging Kepler to finish his work so that his life would not be in vain. Kepler complied but after five hundred years the work of Kepler is still unfinished and if I die without anyone reading my work, it is not only my work that will go unread but also again the work of Kepler as well, will remain unfinished, as it will never be understood in the way the cosmos intended. The Universe introduces a cosmos to Kepler that uses a formula that does not comply with the standards we see in the Universe that we see. Mainstream science work in space and nature works in singularity or time. This might not be clear to any person at this point but that is because mainstream science is not up to scratch even after hundreds of years about understanding Kepler. Tyco Brahe spent a life time accumulating facts and the arrogance of semi blind, self-opinionated, self serving academics in physics of the day that thought they had the authority to decide what science is and what science has to be and what confirms science as much as what conforms science, decided Tyco Brae's studies were pointless.

Up to today the work of Tyco Brahe as well as Kepler is still unfinished and when I try to finish Kepler's work I find that my life's worth of work will be in vain, as modern mainstream physics will see to it that my views will never gets published because I question Newton. I do support Kepler but Newton never supported Kepler and because I support Kepler and not Newton, I am stupid and I am mentally retarded, slightly weak in thinking and mostly out of touch with reality.

But that was how everyone in science at the time also felt about Tyco Brahe, Kepler and Galileo. I feel the same as Tyco Brahe, as Galileo Galilee felt and more, because I am bullied in the same way. Newtonians of the day keep going on about the Church and the wrongdoings committed to Galileo Galilee, but it is modern science that killed Galileo Galilee, because everything Galileo Galilee said Newton turned around and destroyed.

Gravity is no force...gravity is time because that is why one may employ the pendulum arm invented by Galileo Galilee to measure time. Time to science is how long it takes for the earth to circle once around the Sun, but for God sake, how can that time apply to the entire Universe. The pendulum swing you would use in the atmosphere of a massive star to measure the gravity applying will show time very different in the star. The thing about Newtonians is that they are not accountable to no one about what in their world contradicts reality. The one polishes the other and the lot shine while everything underneath is rusted to the core. And there I have lost the audience of every academic in physics, but I guess I never had their pompous attention anyway...

There is no factor such as mass in the Universe. There is no evidence of a factor such as mass that holds any validity throughout the Universe. There is no proof that the Universe indicates the presence of gravity by the measure of mass forming a pulling power and while science conducts an entire religiosity based on this falsified belief, any such notion is falsified truth. **Using science based on the idea that there is a pulling force such as mass forming gravity is as valid as giving Snow White seven dwarfs and then beginning a religion on that basis.** There is a factor such as weight but there is no pulling of anything towards anything by magical forces forming gravity or whatever. **There is a conspiracy of conducting fraud by claiming non- existing forces but such claims are utterly fraudulent.** I have been trying for twelve years to introduce the true forming of gravity but all Physicists I have encountered prevented me of doing so. They stop me because my work makes Newtonian science and when removing the notion that a pulling force of gravity works by the value of mass, most of their work becomes science fiction that falls apart in substance. Read this and see how **students in physics** are methodicly **brainwashed** to get the students to believe in the absolute accuracy of science. Professors and teachers participate knowingly or unknowingly in this thought manipulation process by means of conducting **mind control.** By applying this **mind-altering process** those teaching physics ensure they subdue students into becoming mind-altered zombies. It is a process going on for centuries and which without science would have no foot to stand on in the modern environment. By presenting incorrect, falsified or unproven facts and other untruths as proven truth they exert **thought control** and thereby change the student's ability to appreciate what is correct and believable logic and then force students to discard such judgement ability in favour of accepting the institutionalised untested norms and values of science in order to unequivocally believe in science. The accepted teaching methods force students to comply by compromising their better judgment and then systematicly to capitulate under teacher pressure by making their own what science prescribes what should be believed. I prove this and you get this for free so what have you got to lose...**but you can get wise to what forms a better**

understanding about science! By using the building blocks that forms the Universe I take you back to the instant the Universe started and I show you how the Universe fits like a jigsaw puzzle.

Albert Einstein formulated a concept in 1905 he called **The Special Theory of Relativity** and in 1915 he introduced his assessment on the principle of **The General Theory on Relativity**. I do not quite agree with his findings. What I discovered goes far beyond the discovery that Albert Einstein formulated. I have discovered that the Universe is not employing a general relevance of singularity, but throughout the Universe there is a fixed overall state of *The Absolute Relevancy of Singularity* that is not only **controlling the Universe**, but is what the Universe **constitutes of**...it forms the Universe...it is the Universe. In his conduct Albert Einstein suggested the **Universe goes "flat "** according to his calculations. In **a "flat" Universe** all dynamics of dimensional space would disappear. That would only be the case if and when the three-dimensional space we see would become single dimensional and everything will relate to 1 because after all that is what singularity refers too to being one dimensionally. With all the genius of Albert Einstein and all the magnitude of his vast mathematical experience, he did not take the Universe further and in some quarters even become some sort of laughing stock for not proving what he proved mathematically. In this article the purpose would be firstly to prove he is correct in what he said and secondly to prove he is incorrect in his method of conduct he tried to use and thirdly to show why he was wrong in his approach to use the dynamics of formulated mathematics to try and go there into singularity.

However, notwithstanding the magnitude in significance *The Absolute Relevancy of Singularity* presents as a breakthrough in science, the influential members of the scientific establishment will not recognise my theory on **The Absolute Relevancy of Singularity**. Past encounters taught me that mainstream science in physics will again ignore the ideas that I formulated as *The Absolute Relevancy of Singularity* and I don't believe it would be well received, it will be seriously considered and much less be accepted by those with the authority to change physics principles. In spite of science fondling the idea they know every iota there is to know about science as a subject...

==What you are about to read is as new to the sciences of cosmology as iron was to the Stone Age people.== Because it is a new way a new way of thinking bringing truth to the science of physics I think the theory I introduce would never be accepted by the paternity governing science during my lifetime because science is fixated on Newtonian ideas, on playing games with mathematics. By creating a dream world with mathematics playing God they get into a role which makes them bent on believing in the outrageously marvellous, and the unexplainable magical powers with gravity working by mass supplying a pulling power, which is a fact never proven and accepted only on Newton's word and Newtonian cultural bias, and in the face of all of this they still claim to only use proven facts. What I ask of readers is to beforehand forfeit the culture of Newtonian bias when reading this by paying attention to what I say and not about the degree in which I stray from mainstream science's thinking. This way the exercise will present many new ideas and when explaining my new concept, the new principle will become clear.

There is so much to benefit from. Science has no idea what a Black Hole is while I can prove what a Black Hole is. I formulate mathematically what ==**"the sound barrier"**== is. I prove what gravity is. By using the four cosmic phenomena, which is what the cosmos uses to form gravity, I show what ==**"the sound barrier"**== is and I go much further than that. I show that gravity forms from using the **Roche limit**, the **Lagrangian system**, the **Titius Bode law** and the **Coanda effect**. I uncover these principles by placing Π within the formulating of gravity and when using Π I bring clarity to the misunderstood cosmic principles. The list of the unknowns I can then explain is almost endless. ==**Gravity forms by movement that establishes singularity initiating a circle in using Π.**== I show why gravity is there, how gravity forms and what role stars play in forming gravity. There is no difference between how gravity and electricity forms and that I prove mathematically by decoding the cosmos.

I prove mathematically when atoms spin they establish Π that forms the Universe. Whatever forms gravity, that has to link closely to Π since everything that has anything to do with gravity forms a circle that is Π by the value of the square radius. If mass has anything to do with generating gravity, then mass has to apply Π or otherwise mass has nothing to do with the forming of gravity. Everything using gravity forms a circle of sorts, which forms the curvature of space-time, which is Π and which curves light. The way the planets orbit the Sun and how stars spin has all to do with Π. In spinning in a circle, Π forms gravity as a centrifugal force that condenses space.

While I prove nature, Newtonian science cheats, corrupts and manipulate nature to make science work in ways nature doesn't work. That statement I prove and I prove undoubtedly there is no pulling of any object on any other object. Gravity does not work like a magnetic field but works in principle by the object rotating… But it is space that compresses as rotation reduces in a principle called the Coanda effect. It works by material objects producing movement in rotating condensing space around the object of to liquid. Gravity turns space surrounding the rotating material from gas to freeze it into liquid.

If you ask me what is the difference between how I see gravity work, which I can prove by going to nature and how Newton's gravity works, which unbelievable as it is but I also prove has never been proved by science it is nature.

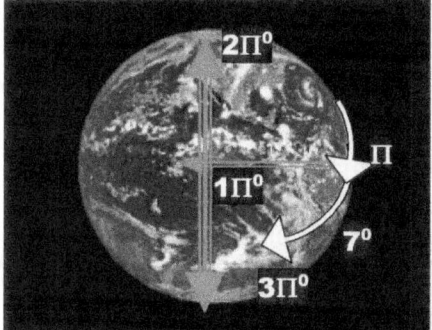

The entering by singularity that goes from $7°$ by $3\Pi^2$.

The extending of singularity goes from Π^0 to $5\Pi^0$.

I do that while I also show that Nature does not use Newton in any way that is what makes the difference. Newton says objects pull by an unexplainable magical force called gravity while I say as round objects rotate it forms Π and by collapsing Π the space around the star / planet that space compresses and thereby collapses from $21.991 / 7$ to $3.1416 / 1$.

As it turns it divides 7 from $21.991 / 7$ by 7 to form $7/7 = 1$. As the direction rotate Π changes the travel by $7°$ and that $7°$ is $7 / 7 = 1$ or $\Pi = 3.1416 / 1$.

Then by getting reduced it compress everything in that space as the entire space the object holds and claims condenses. Space reduces and Newton said objects in space pull each other.

This compressing of space is the Coanda effect. The difference between what my approach must be when using nature and Newtonian's approach that is unsupported by nature is one Universe away from each other. I show a functional Universe and Newton show what they call the mysteries of science in the Universe. I prove everything that Newtonian science this far could not prove ever. I prove the Universe applies four keys by which gravity works instead of unexplainable magical forces that pull each other and this no person (not even Einstein) this far could ever There are 4 principles applying gravity a value of Π as a circle rotates.

By implementing the 4 cosmic principles I book thereby I show how nature works keys while I show how science falsifies make science seems to work by magical pulling. explain. and forms show in the with the 4 facts to forces

What I show was never printed before by any person and is understood for the first time in human memory. These principles redistribute cosmic density from liquids to solids.

What you read from now on is completely new and I am the first mind on earth to put the information you are going to encounter in print. Should you not agree with my statements then purchase the book **_Uncovering Corrupt Science_**. If you read and you wish to confront me do so after you understand the proof I supply in **_Uncovering Corrupt Science_** and I would gladly explain that part which is not clear to you. However start where you are comfortable with the information you read because understanding science is most important!

1 ALL THERE IS... ABOUT NOTHING?

I wish to share with you my very first article I wrote when I still penned down thoughts about what bothered me about science. I never for a second at the time envisaged me writing about science but I had an urge to sort out what I couldn't understand. It was at a time when I still held Newtonian science in high regards and thought it was I that had some problem with expressing ideas I could not understand. At the time I still had an urge to understand what others said they understand before I came to understand Newtonians understand **"NOTHING"** about the **"NOTHING"** I can't understand.

The single most tedious problem I faced in writing any book was where to start. This book is intended to discuss everything about the **Universe that is** in time and in space, which on its own is eternal and put together, remains eternal. Wherever I decided to begin there were always some factors that lead to that event. Firstly, I had to explain those factors that the readers were to understand, before I could explain the events that took place wherever I decided to begin. Unfortunately, these factors did not stand alone, but were supporting other events that I first had to explain. If there were not a well founded comprehension about the factors that led to the events which supported the factors in explaining the events what occurred before the start of whatever point I decided to start off with, it seemed as senseless as this sentence which I just wrote. Writing one sentence, while sounding stupid is one matter, but to write a whole book and coming over sounding like an idiot is quite a different kettle of fish to fry.

That's what they have gone and put in the Universe...nothing! Those Newtonian idiots went and filled everything in the Universe with, nothing. They filled the Universe as far as the eye can see and much further than what a telescope could see with nothing. I had one aaaaaa and that is that I couldn't be a bigger fool than the fools calling their positions as experts in cosmology! To fill a Universe with nothing... how expertly bright and intellectually superior can that be...even if you are a Newtonian mathematician?

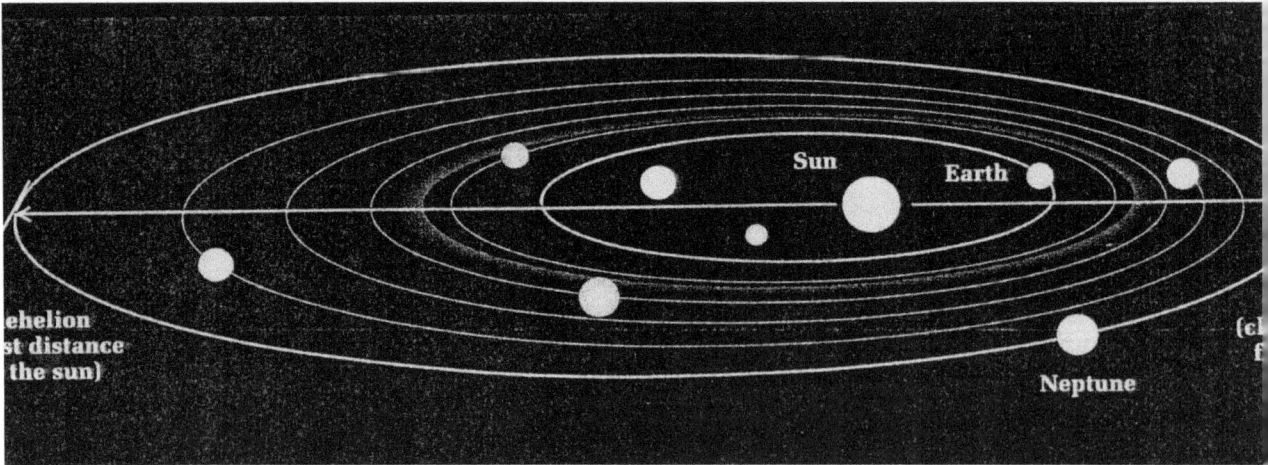

The obvious point to start with, should be the Big Bang, with one minor problem. In the Big Bang, the Universe was not a **"NOTHING"** that came from "nowhere". The Universe is forever **"SOMETHING"** which is flowing from "somewhere" to another destiny which seems to be right in the middle of wherever the Universe is heading. That was what forced me to start with the **"NOTHING"**, and only by starting with **"NOTHING"**, I stood a chance in not achieving **"NOTHING"**.

I feel obliged to explain the stance of **"NOTHING"** and my criticism about **"NOTHING"**, You might think I am writing about **"NOTHING"** but this very **"NOTHING"** is what forms most of the **"NOTHING"** that is all the space in outer space according to Newtonian intellect Newtonians think that between the moon and the earth there is **"NOTHING"**. The way Newtonians also think is that between the sun and Mercury there is lots of **"NOTHING"**. Would it then surprise you to know that Newtonians think that between the sun and Pluto there is lots and lots and lots of **"NOTHING"** From the earth to the moon the Apollo space mission travelled through lots and lots of **"NOTHING"** and the **"NOTHING"** stretched all the way up to and even further than the back of the moon because they found **"NOTHING"** even there. It is important to know that on the moon surface there is **"NOTHING"** and on the back of the moon there still is **"NOTHING"** but that **"NOTHING"** that is on the surface of the moon is very much different from the **"NOTHING"** that the Apollo missions went through on their way to the moon. Between every planet and the sun the **"NOTHING"** is escalating to the point that the **"NOTHING"** is doubling between every planet that is separated by **"NOTHING"** and this doubling of **"NOTHING"** is continuing beyond Pluto where there is even more **"NOTHING"**!

This was my first problem I encountered with Newtonian cosmology. If they were on the moon and found **"NOTHING"** and they travelled all the way to the moon through the **"NOTHING"** in outer space did they encounter **"NOTHING"** in outer space that was comparable to the **"NOTHING"** they found on the moon and where then according to their most informed opinion can we see the most **"NOTHING"**. Was there more **"NOTHING"** in space or on the moon? This was a big problem I had when I first started to digest cosmology and as I grew so did my confusion grew about everything that was **"NOTHING"** in outer space.

In order to understand the Universe you had to understand Einstein's theory. Einstein said there are three known substances in the Universe that we know. One is matter, two is time and three is space. Matter is one substance, which we consider everything should consist of; time is a **"NOTHING"** that man created and space is a **"NOTHING"** that God created. Newtonians fill everything forming outer space with **"NOTHING"** and according to Hubble and his constant; this **"NOTHING"** is even getting more!

However, have you, as the reader, ever considered what **"NOTHING"** is? What is **"NOTHING"**? It is a notion used by all and understood by none. No person can define the exact meaning of this word. In every day language, it applies to an understanding of not understanding. This seems to apply to the world of cosmology as well. The Universe takes its birth from this very word. However, there must be some explanation to the concept of the meaning of the word, if it has any right to exist.

Only when one put into context with its contrary meaning, which is **"SOMETHING"** there comes validity to the meaning of **"NOTHING"** and only then can one visualize the concept of **"NOTHING"**. Let us use an imaginary scenario. Let us pretend it is nighttime and there is a dog outside that starts barking. The owner of the house gets uneasy and goes out to investigate. According to the dogs, observation there was **"SOMETHING"** to investigate. However, to the person's observations there was **"NOTHING"** outside. Who is right?

Let us take this same scene a little further. Now the dog enters the house. In the living room, he sees humans staring at a light that flows from a square apparatus. The people's faces seem taut and hypnotized. Because of the dog's inability to see the picture in the T.V., the dog will find the human behaviour confusing. To him, they focus on, **"NOTHING"** but to their vision's ability, **"SOMETHING"** excites them.

With this in mind, that the concept of **"NOTHING"** can only be defined by the use of the word **"SOMETHING"** and in this doing, **"SOMETHING"** validates the meaning of **"NOTHING"**. With the next verbal sketch that might seem to be a little tedious, I would like to prove my point. I use myself as an example as not to gossip about the facts of another person.

There are parties in my family who thinks of me as **"NOTHING"**. According to those persons, my outlook on life is flat, boorish, single minded, opinionated, mindless and rude because of my personal standings and viewpoints.

According to me, I consider myself to be to the point and without frills. I am the owner of a B.M.W. 735 I but the model is a 1986, thus almost 16 years old. It has little value and nobody is interested in buying it, though it was an expensive German car.

I live in an ordinary farmhouse that is far from good looking. I own a farm with an enormous debt. My five tractors, two centre pivots, three trucks and all my implements are between 10 and 30 years old. According to my standards, these antiques that the bank regards, as

"NOTHING" is all I have. That means I am financially worth "SOMETHING" because it is better than having "NOTHING".

My relatives, seen from my point of view are gaudy overindulging with liberal notions because they benefit financially from being "liberated". To my mind, they hide their shortcomings behind a curtain of money and therefore their outlook on life is "NOTHING". They consider me as "NOTHING". I consider them as "NOTHING". Who would be "NOTHING" because we can't both be "NOTHING".

We cannot both be "NOTHING" because then "NOTHING" must be sharing in a common and specific value. If this was true, we then must have had "SOMETHING" in common though it will be "NOTHING", but we have "NOTHING" in common therefore we share "NOTHING". We have "NOTHING" in common. With all this personal gossip, my only intention is to give meaning to the "NOTHING" that I share with my relatives and not even that "NOTHING" is "SOMETHING".

This brings about that "NOTHING" only has meaning when compared to "SOMETHING" which then can put a certain value to "something "and rob "NOTHING" of every value it might have.

This misconception divides families political convictions, religious believes, and cuts through society to the bone. In addition, all of this comes about because of "NOTHING". Every race thinks the other race is "NOTHING". Every country think the other country is "NOTHING". Every religion think the other religion is "NOTHING" and therefore wish to convert the other party to this person's religion whatever the religion might be.

Let us take this "NOTHING" even further. As a conc in the Universe, we are able to take this "NOTHING" much further. The atom compromises a nucleus in th middle and electrons orbiting the nucleus. In a sketcl paper this two dimensional drawing seems exception and the atom seems to be logic.

However, if you put the atom in the correct perspective, your comprehension of "NOTHING" becomes mesmerized, stripped of all logic even to a point of absolute inconceivability. To prove this point I will place the atom to a more lifelike scale.

Put an object the size of a sugar grain in the middle of a rugby field in the very centre. Consider this the nucleus of a hydrogen atom. Now put a grain of salt right in the middle of the upright post, fifty yards away from the grain of sugar. You now have constructed your hydrogen atom. The orbit of the grain of salt (electron) will now move in a circular motion around the grain of sugar at the speed of light from the one post to the other post.

Now go and stand on the pavilion at any given angle and admire your hydrogen atom. Your atom would be unsighted to yourself or any other person with normal eyesight. Remember the two grains are representing the "SOMETHING" and the grass is representing the "NOTHING" part. The grain of sugar and the grain of salt is the portion that one can regard as the matter part and the grass represents the "NOTHING" in the atom. There seems to be a very little of "SOMETHING" and a tremendous amount of "NOTHING" in this picture.

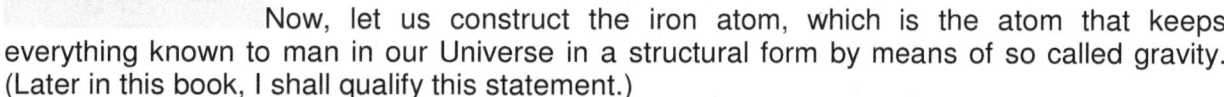

Now, let us construct the iron atom, which is the atom that keeps everything known to man in our Universe in a structural form by means of so called gravity. (Later in this book, I shall qualify this statement.)

For this experiment, you would have to go out to a vast expansion like the Namib dessert. Choose any point and put down a golf ball. Now put two grape pips opposite each side of the golf ball at a radius of 10km. Now in a radius of 20 km from the golf ball 40 km in diameter put down eight grape pips evenly apart. Now place another 16 grape pips in a diametric circle of 80 km (radius 40 km) evenly spread over the circle.

If you want to construct an even denser atom, you would have to go another 20 km and construct a circle with 32 pips in this circle. However, let us stick to the iron atom. Now you would have to fly in an air balloon at a height of about 500 km to picture your construction of one of the densest atoms in the universe, which make up stars. Even if you had all the imagination of a space scientist, (and do they have over inflated imaginations!) you still would only be able to see sand. Now this represents an atom of enormous density! The sand in fact represents the **"NOTHING"** that this enormous dense atom is made of. This boils down to an enormous misconception that overruns any sort of sane logic and it is no wonder that none of the "SUPER- EDUCATED- MASRER- OF- FACT" and most learned cosmologist experts ever tries to explain the workings of this **"SOMETHING"** to **"NOTHING"** ratio.

From this stance the **"NOTHING"** is so overwhelming that the **"NOTHING"** to the **"SOMETHING"** ratio becomes unexplainably unreal. If the **"NOTHING"** in this argument is so overwhelming, the only conclusion I can come to is that that **"NOTHING"** must be **"SOMETHING"**.

In a question I once put to a "SUPER- EDUCATED- MASTER- OF- FACT" Genius, how the construction of this atom can bring about a dense and solid structure as in the case of iron, he started blabbering about the electrons moving so fast that it forces the atom to become impenetrable. The worst part was that I had to sit and listen to this rubbish. What happens at 3 000 degrees when iron becomes a gas. Does the electrons then become slower as it orbits the nucleus?

This means that I have to conclude that the speed, at which the electrons travel, will bring about whether matter is solid, a liquid or a gas. This means the negative particles determine the density of the atom. This is not the end of the argument.

Now, picture the scene where the electrons orbit the nucleus of the atom, in an orbit that place them right next to the nucleus. The electrons are now rubbing against the nucleus in such a way that **"NOTHING"** separates the two. Will there be more **"NOTHING"** between the electron and the nucleus in the last mentioned relation or will there be more **"NOTHING"** in the previously mentioned example. In which of these two examples are there more **"NOTHING"** that separates the electron and the nucleus of the atom? Whatever your answer may be, then why? I do not care how brilliant you may think you are, but to this question there can be no answer as to the value of **"NOTHING"**.

I believe predominant in my Creator. That includes my belief that His creation is perfect. If the creation were perfect, there would obviously be logical, calculated and obvious solutions in the composition of the universal creation. With this construction of the atom, science obviously fails in being calculated and logic. There is far too much **"NOTHING"** and far too little **"SOMETHING"** to make any sense. In this lies my main motivation why I conducted my studies, because I believe that science and religion should be the very same thing if the same Creator creates it.

The atheists believe in **"NOTHING"** after death. When he breathes his last breath then according to him, there is **"NOTHING"** that follows this departure of life. Those same atheists are very quick to point out how valued life is and how dear we must consider it to be.

No atheist, up to now could explain to me the reason why, if life is so precious, would it then just be lost after death to **"NOTHING"**? Why then would **"SOMETHING"** which is that precious (according to the humanists), just vanish into **"NOTHING"**?

Personally, I think that all atheists are on a self-sublimation crusade. Their way of thinking is that if man, including the atheists are without a god, man then is put right on top of the priority list and becomes a god! As scientists, they become even more of a **"SOMETHING"**, because the scientist becomes one of only a few that can understand and explain to a very trifling point how creation actually works.

However, if there is a Supreme Being, then this blasphemer and self appointed god becomes just another **"NOTHING"** as all the other "nothings" that call themselves humans. The atheist tumbles down to the same level that we, the "nothings" find ourselves to be in the order of importance. Then the atheist is not a **"SOMETHING"** any longer and he could not consider himself one. By recognizing a Supreme Being, the atheist will become the same **"NOTHING"** that he regards the rest of us to be and that would be the end of his sublimation in regarding himself to be **"SOMETHING"**.

UNCOVERING CORRUPT SCIENCE

Part 2

You are going to read about a conspiracy but people think of a conspiracy in many terms. Let us define not by definition but by interpretation to what a conspiracy constitutes. What do you think is a conspiracy?

All the conspiracies you know about is known about because someone somewhere makes money by allowing the revealing of the conspiracy. Silencing the conspiracy does not make money but informing a suspicious public loosens the flow of money. If it were a true conspiracy no one would know about the conspiracy because the powerful would make money from not revealing the conspiracy. The revealing of the facts about any conspiracy would be stopped before it leaked because it would kill the flow of money.

A conspiracy is thought to be a gossip story that makes money and by not revealing it or revealing it goes in line with making money or not making money. You can download this book free of charge because I don't make money by revealing the conspiracy. I truly want to find an audience to divulge the truth. I want to make money but it is by showing how I can correct the flaws in science, not by hiding it in a conspiracy.

People put a conspiracy in the same realms as a gossip story, an old wives tail, which is going about but does not intend to harm and mostly serves as amusement to many. Hearing about a conspiracy tests your intellectual comprehension. It is some quiz that you match your truth against the truth that the conspiracy reveals. It is a funny, but it is not funny until you catch the funny part hiding behind the conspiracy and only when you measure the catch behind the conspiracy are you treated to be amused.

If the conspiracy does not touch the person directly then no harm is felt and no harm is intended. Every one holds this view that a conspiracy is on a slightly higher level than gossip. It is a gossip story about someone living in the neighbouring village known only to some people next door but has no direct linking to me or has no threat to the safety of others directly associated to me. Everyone treats a conspiracy as if it is something amusing that holds no threat at all. It is something that goes around as a joke of sorts.

For all of those that lives in America or on the far side of the Moon or spent the past three years on a vacation on Mars and others that had no contact with Mother earth or the media on Earth, this is Madeleine McCann the daughter of two prominent Scottish doctors that was on vacation in Portugal in 2007. Since her abduction no one saw her, no one heard of her and no one thinks of her. However that is not new because since the time of her disappearance more than 300 000 other kids in Britain alone befell the same fait and there is no media-mania to try and find those children.

Lets narrow the conspiracy idea down to a known case. A conspiracy is very typical to the Madeleine McCann case where the disappearance of Madeleine brought money to many. Madeleine McCann disappeared Thursday, 3 May 2007 and was / is the daughter of Gerry and Kate McCann. Anyone living in the western world except Americans would know about Madeleine McCann. Americans live in a unique Universe called "the States" and regard Europe as some overseas country situated just past the moon. Americans are the most uninformed brainwashed nation on earth thanks to advertising. Americans can't cross the street without some advertising jingle telling them how to do it the best way while getting the most pleasure while crossing the road and for fun they will throw in a sexual connection in the crossing.

Madeleine disappeared from a hotel room in Portugal while being in bed. For months the disappearance of Madeleine was front-page news and we heard about her all day long. We heard about Madeleine's disappearing day and night on all channels and in all papers around the world. Then the news raised the hopes and then the news dashed the hopes but it was all about the disappearance of a girl called Madeleine McCann. To sell advertising time the press had to write about Madeleine.

This is a conspiracy.

Not the fact that she disappeared or didn't disappear or that she came to harm or died at the hands of her mother or not. The conspiracy hides in big deal that was made about the case all throughout the months afterwards. The case was blown out of all context and proportions considered the international effect it had. The press made an issue about the case as if it was some special occurrence unknown to mankind. The Press presenting this case as a one-off incident makes the entire case a shameful mockery to society. The world viewed this, as an incident equal to the Lindbergh baby's disappearance in uniqueness while the truth is that 11.4 children disappears hourly in Britain alone and no cockerel ever crows about it.

Britain alone loses a hundred thousand children under the age of sixteen per year every year. One hundred thousand children go missing each year in Britain alone and what Special Forces are in place to combat this...nothing. The government never mentions this in parliament. In 2007 there were a number of bomb blasts on trains and one bus blew sky high and that brought Britain to a stand still. The number of lives lost was many times less than one hundred thousand and such bombing happens once in about ten to twenty years. Yet, the propaganda values in terrorists out ways the leverage of disappearing children.

Under tough British laws, anyone of 30,000 on sex register, those marked as sex criminals must inform police when planning to travel abroad. That information has been used to compile names of at least 130 paedophiles known to be on Portugal's Algarve coast, at the time when Madeleine McCann vanished. Police searching for the missing Madeleine McCann were as they claimed scrutinising four pieces of "very useful" fresh information. The tip-offs came among hundreds of calls made from Portugal via a special UK charity number. If these figures were terrorist-related incidents the world would be in world war 3 by now.

This is not all. It is said that each year, non-family members, often in connection with another crime, abduct more than 58,000 U.S. children. Family members who are seeking to interfere with a parent's custodial or visitation rights abduct more than 200,000 children. Although the vast majority of children (at least 98%) return from abductions, too many children do not. While there are only around 100 reported cases each year of the most dangerous type of abduction – stranger kidnapping – fully 40% of these children are murdered. Where is the media coverage on these events? According to the Department of Justice, almost 800,000 children are reported missing to law enforcement each year, while another 500,000 children go missing without being reported to authorities. I was astonished to find recently that 1 million children, yes 1 million, go missing every year in the US and the UK. These are astounding statistics, every year a city the size of Amsterdam goes missing? Less than 200 of these are murdered so where are the rest? Its mind-boggling stuff, when you delve a little deeper, in the UK a child goes missing every 5 minutes, 2000 kids a day lost in the US! I suspect hat even these figures are suppressed. Getting excited about these figures still hides the overall truth and we will never get to the real numbers.

Belgium is a small country of some ten million people and yet in the Brussels region alone it was revealed that 1,300 minors disappeared between 1991 and 1996. That was only the reported cases and not those that went unreported which are the cases we will never know about. Around the world you find the same repeating story and multi-millions of children go missing every year never to be found. This is not even contained to a continent or a part of the world but is Global. There are children going missing in every country across the world. We dare not mention numbers when we are thinking about what is happening in countries such as Brazil, India, China, Russia and other countries where due to the geographical vastness the Governments just don't have any records of missing persons in place. The most remarkable issue is the absolute silence we experience about this matter. It's as if its not for real and we know it is. I can't confirm what I repeat from the Internet but if it is true it confirms a lot of my suspicions I have about this. Where do they go? It is millions of children in the end. The numbers alone just can't hide the crime but the apathy of all the Governments can apparently hide the numbers. It is another conspiracy kept in a cloak of silence. There is apparently a program or was to be a program on TV called Conspiracy of silence, which is / was a documentary listed for viewing in TV Guide Magazine and that was to be aired on the Discovery Channel, on May 3, 1994. I have seen on the Internet that this documentary exposed a network of religious leaders and Washington politicians who flew children to Washington D.C. for a sex orgy. I can't confirm this that I got from the Internet but my logic tells me not to deny this either. There is a show that was supposed to be aired about this disappearing of children. At the last minute before airing, unknown congressmen threatened the TV Cable industry with restrictive legislation if this documentary was aired. Almost immediately, unknown persons who had ordered all copies destroyed purchased the rights to the documentary...Only the most powerful on earth can fasten a lid on the matter this secure.

Nearly 30 million children and youth go online to research homework assignments and to learn about the world they live in. Research by the University of New Hampshire found that one in five children between the ages of 10 and 17 received a sexual solicitation over the Internet during any given year. One in thirty-three received an aggressive solicitation - a solicitor who asked to meet them somewhere; called them on the telephone; or sent them regular mail, money, or gifts. This is very difficult to comprehend, we are not talking of two or third world country's here, wouldn't you think the media would be all over this for instance? Or our respective governments constantly warning us but no, I only managed to stumble on to these unbelievable statistics on an obscure website. Belgian parents live in fear of paedophiles because it is very well known that in Belgium more than any other place these paedophile rings are almost openly active! When I hear about doubt or I hear people say that if child abuse, kidnap and murder were happening on the scale it is suggested to occur then it would not be able to be kept under wraps. The child kidnappings and murders in Belgium in the mid-1990s threatened to implicate the country's political establishment and other famous names. That power that democracy places in the hands of politicians is where the power is that control absolute silence while the banks control the political power by telling the politicians what the bankers wish to happen and the crime syndicates control the banks by telling the banks what it is that they wish to happen. Therefore democracy obeys the rich that controls then banks that orders the politicians that rule us. What happened in Belgium shows that the syndicates can control the people. This is the only explanation as to how they got away with it and pinned it all on a sick and pathetic paedophile and child-supplier called Marc Dutroux. It was the classic establishment response to danger that is repeated constantly across the world. As so many have said so many times, people generally don't get their views and opinions from researched information, but from an 'image', an 'impression', of how things are. By controlling the leaking and presenting of conspiracies then by controlling the conspiracy they control the information leaked as a conspiracy. With regard to missing children, this 'impression' is heavily influenced by the number of lost children stories they see in the media. The Belgians live in fear of having their children abducted, as the paedophile gangs operate unbothered by the Police. Once a child is abducted, they are never seen again. The Marc Dutroux case became famous, as he allowed girls to die in his cellars when he was arrested. But two girls escaped unharmed and provided many leads as to who Dutroux' clients would have been. But no inquiries were made. Belgians all have their theories as to who is involved, many suggesting paedophiles are high up in the government. In Brussels, kids are often photographed at play and abductions made to order. Why the Police don't do anything, or politicians keep silent is not known for sure, but you can use your imagination...a combination of threat and reward - in a country where money easily buys influence, it is not surprising. The strain of knowing so much evil and not being able to act to prevent it gets too much for many. Belgian Police are very prone to committing suicide. The Sun recently reported that Algarve is 'haven' for paedophiles. Why must society's parents tolerate these criminals scum living amongst us?

In Britain, Missing People -- a charity formerly called the National Missing Persons Helpline -- has tried to draw some degree of attention to these thousands upon thousands of missing cases. It is astonishing to witness the degree or actually the lack thereof to which the news media devote attention to vulnerable missing persons. One little undefended organisation is claiming that despite its efforts to generate news coverage for all missing persons cases, the news media themselves will cover only those a very modest few cases which the media claim that those cases fit their publications. The Media should eat this storey as if it was a ripe banana and yet they don't. Why would that be, why would they ignore the bulk of this storey. If one or two cases could generate such a massive response and earn that much advertising revenue, what prevents the media to go all out on a mission to bring justice to those that went missing.

Two cases of missing white woman syndrome are given as contrasting examples: the murder of Hannah Williams and the murder of Danielle Jones. Although both victims were white female teenagers, Jones received more coverage than Williams. It is suggested that this is because Jones was a middle-class schoolgirl, whilst Williams was from a working-class background with a stud in her nose and estranged parents. Media reports about the murder of Amanda Dowler, the murder of Sarah Payne, and the Soham murders as examples of "eminently newsworthy stories" about girls from "respectable" middle-class families and backgrounds whose parents used the news media effectively. These cases are controversially, that in contrast, to the street murder of gangland murders by youths on youths in drug related turf war incidents which receives little news coverage, with reports initially concentrating upon street crime levels and community policing, and largely ignoring the victim. The assertion is that "the near hysterical outpourings of anger and sadness that accompanied the deaths of Sarah, Milly, Holly, and Jessica" brings ratings to news events. The National Centre for Missing Adults has also commented on the phenomenon by saying "Unless it's a pretty girl aged 20 to 35, the media exposure is just not there. Mentioning these few names fall much short of the numbers of children disappearing pointing fingers to the press. Some 2,300 Americans are reported missing every day, including both adults and children.

But only a small proportion of those are stereotypical abductions or kidnappings by a stranger. For example, the federal government counted 840,279 missing persons cases in 2001. All but about 50,000 were juveniles, classified as anyone younger than 18. About half of the roughly 800,000 missing juvenile cases in 2001 involved runaways, and another 200,000 were classified as family abductions related to domestic or custody disputes. Only about 100 missing-child reports each year fit the profile of a stereotypical abduction by a stranger or vague acquaintance. Two-thirds of those victims are ages 12 to 17, and among those eight out of 10 are white females, according to a Justice Department study. Nearly 90 percent of the abductors are men, and they sexually assault their victims in half of the cases. To further complicate categorization of cases, the FBI designates some missing-person incidents—both adult and juvenile—that seem most dire as "endangered" or "involuntary." Kim Pasqualini, president of the National Centre for Missing Adults, said the media tends to focus on "damsels in distress" — typically, affluent young white women and teenagers. The media's dilemma is that government research shows that victims of non-family abductions and stereotypical kidnappings are most at risk of injury, sexual assault or death. "Damsel" cases may be the exception, but they often are the most urgent. These numbers are staggering and yet even so they are ultimately never mentioned because? Compare what is mentioned as cases to what the cases are mentioned and are targeted by the press to what disappears in reality. What makes the media so reluctant to tell to entire story and blast information out for months?

That is the media response of what happens to losses accruing in terrorist cases but the numbers of losses hardly match. In relevancy one could say for every two million children lost to prostitution there are five bomb blasts. During about the same time there was a huge bank robbery and the culprits were traced far and wide. They never stopped searching until everyone was caught and every note was accounted for. The leader was traced back to an Arab country (I forgot the name) and was brought back to face justice. He was chased until he was caught showing that criminal cases can be solved, except where children disappear. If one bank robbery is committed they get the thug but when children disappear who cares? If we are talking the national media there are very few cases reported in the light of what takes place. The enormous coverage of the missing British girl, Madeleine McCann, who was abducted while on holiday in Portugal, is a rare example compared with the number of children who disappear every year. While people get their impression of scale from the lost children featured in the media, staggering numbers of children go missing never to be seen again. I remember calling many American states a few years ago to ask for their missing children figures and it was truly extraordinary. On average, around 3,000 children a day are reported missing in the United States, never mind those the authorities never hear about. Add them together and you are talking hundreds of thousands of children.

Yet in that time it took to close the bank robbery case about 200 000 children under the age of sixteen went missing and the number that disappeared were never even reported! If there is another two major BIG bank robberies in the same period no stone is left untouched to solve the problem of banks robbed and "terrorist" attacks against "the people of Britain" but to lose a million children every decade or so is quite acceptable because the disappearing is part of a conspiracy and true conspiracies are never revealed or investigated. If a bomb blast occurs in London MI 6 forms a special investigating unit but no Special-Force action is ever taken to bring this loss of children into the open. Losing 100 000 of the prime persons, the future of the country is a matter of discussion that is never discussed in polite conversations. It would be very embarrassing to ask the prime Minister or President of America what happens to 100 000 children under the age of sixteen in Britain alone that disappears every year. This applies to all countries. Which British politician ever made it a political pledge to search into this crime? No one mentions it because it is best wiped under the carpet. Most of the children are a product of broken homes and pupils of the unmentionable side of society and therefore no one cares to care for them. Their parents never vote because they are mostly the driftwood that never vote and the children are too young to vote so who cares? But when two medical doctor's child disappears and they know how to manipulate the press the world goes fanatic. The couple even found an audience with the Pope no less. This has to do with money. The fact that 100 00 children disappears while it never forms a political debate alone tells the whole story. If Bankers make money Politicians have to hush about the Bankers' very rich and therefore very influential customers. The crime cartels tell Bankers to tell the Politicians to be quit. The crime bosses are in charge of the Bankers that are in charge of Politicians that make laws we obey.

According to a Portuguese lawyer that worked in the Madeleine McCann case, him being professionally experienced in working with several English clients, any parent being accused of abandoning the children to danger, is a crime under British law that is severely punished by UK laws. Yea, sure and the moon is made of cheese! That is so typical of a conspiracy. There are laws in place that are laws never attended to or adhered to and nobody ever obeys the laws because they are as good as being non-existent. Every law that is in place serves to protect the rich and the powerful by denying the poor and the defenceless any say in Government law forming. Yes they show democracy but they create mass hysteria instead. The Members of Parliament put conspiracy-serving laws in place that serves nothing anyway. TV hosts such as David Frost claim their diligence and their courage and their tenacity. Why don't TV talk show hosts ask any politician about the children disappearing, because the TV station would not permit such questioning? Why do all members of the printing press or electronic media ignore the subject as if it does not exist? Why not attack all the Political Party leaders on a live debate or cross-question them? Why don't the press report every missing child in the same manner as they reported Madeleine McCann? Is it because there then will be no space left to report anything else or who stops them shouting about it?

These children just disappear from the face of the earth and that says there are mighty powerful money barons at work. Those that bought those that have the political power to power to decide who goes to war and die and who gets rich from the declaration of the war have the leverage in the social structure to allow 100 000 children in Britain alone to go missing and the public is none the wiser and could care even less about the matter. Those children disappearing are hard currency more valuable than money. Those children become assets as good as gold or property because their currency is set fast. I believe there is more dope bought with children earmarked for prostitution than there is money used as transaction payment. I believe a lot of oil is bought with pretty little blond haired girls that have gone missing. Telling everybody about the McCann girl is not the conspiracy but not telling the world at the same time about the other 100 000 children that also disappeared becomes the conspiracy. It is not the information that is presented but it is the information withheld from the presentation that becomes the conspiracy. It is screaming from the mountaintops about one girl while pushing the other 100 000 that went missing under the carpet, that becomes the conspiracy. To defer the attention from the true problem they cry about one.

Reading about such a pretty girl being abducted or going lost is sensational. It is something to cry your heart out about. Knowing about 100 000 children being abducted or going lost is a problem and while everyone is ready for the sensation no one wants to know about a bloody problem no one cares about. The following is a joke coming from the Internet:
What's the difference between Madeleine McCann and Elvis?
The Answer: More people believe Elvis is still alive. This is no joke.

To hide this idea of Madeleine being dead is the centre part of the conspiracy, because only then can the McCann family be in the position to earn money through engaging sympathy.

A Portuguese Polices officer spearheading the investigation wrote a book stating his (and the official police) point of view on the case and tells about why they think there is no more evidence to research since they have the opinion that the girl is already dead. He said Madeleine is deceased and gives his reasons in a book. The McCann family got a court interdict stopping his book being published because they want to stop the idea spreading that Madeleine is dead. Why would they spend hundreds of thousands of Madeleine's money to prevent a detective from putting to print his views on the case?

It is about money as every aspect of this conspiracy is about earning money. A dead Madeleine would not enlist donations and evoke the media ratings that the story generate and the income of the entire enterprise will plummet if the girl's death is accepted by the public at large, while searching for a live girl brings in millions and to hell with free speech and free opinions when money is the issue. That is why the papers that printed the fact that Madeleine is dead was sewed and they paid up. To hell with free speech because the papers saw their earning in revenue go down the toilet by allowing the public to think Madeleine is dead. Admitting to Madeleine's death will kill the money flowing in and that is horrific.

Even the money the McCann family and all other people fight to accumulate is a conspiracy. One group of person's thought of as Bankers bought from crooked politicians worldwide the privilege to print paper and give the paper a value. In the system I take the paper they printed and then I have to regard that paper as having more worth than say my home has because I "sell" my home by detaching my ownership from me and attach my own worth to the printed-paper. Then I am very impressed with myself afterwards in my effort to exchange what has visible and useful worth for some paper they tell me I have to accept the worth but has no worth but to give it back to bankers to put in their vaults. This is insane stupidity and every person on earth goes along with the madness. The Bankers decide the worth of the worthless they print at about no cost to them and then by creating a system I am forced to accept their paper and the worth they attach to it or ells I starve and die. Even committing to dying requires money to accomplish the process. If you come to the end of your life you have to pay for the privilege to die and go on.

The Bankers take their tax or share long before the Government can but they call it bank fees. People are so brainwashed and beaten to a pulp by systematic control of the mind and their thoughts that they fall into the practise of doomed slavery without trying to fight for freedom. In the days of the Romans and the Greeks slaves were paid 10 %of what their Masters earned from their services while the Masters still had to feed, cloth and shelter them at the cost of the Masters. We all are slaves to the bankers but we earn about 10% of the cut they take.

While the Mammonites pay us 10 % of what we earn from what they earn from or services, we have to cloth ourselves, feed ourselves and our children while we purchase houses form the Mammonites and then find that behind successful Mammonites there are Bankers pulling strings by supplying worthless printed paper we accept as the commodity we will work all our lives to accumulate and possess. In the end we can't take with us anything we ever wanted on earth because it is worthless.

From every angle this case presents including the Pope's visit all aspects involve money and it is about publicity that will entice donation of money leaving very little scope for a girl being found. Why don't McCann donate some of the money they have to searching for all children that has gone missing in the time Madeleine went missing? Why don't they also include as many photographs of children that went missing during the time that Madeleine went missing? Because then it would not bring about the money that it makes when only one very special little girl is in the hands of a child molester…but 100 000 children just vanishing slips the minds of every "concerned" do-gooder in England while they donate money to this deserving quest just to keep their minds away from the 100 000 others. This tendency to evade the truth is a sickness that runs through all aspects of society because this disease is what we use as education. I am going to show how teachers in science **brainwash** students by forceful **mind control**.

Where and when did it become acceptable and even come across as being intelligent when by behaving intelligent you lied and cheated and doing it with cunning then be thought of as a wise and clever person? This is when those that showed much criminal intent and unsavoury behaviour introduced a god that is dead and that is the god "money". We all bend in front of money and we all crawl before money and money is a god that is dead. On earth we fight for one god who is money and the god money finally destroys every person. When I say I am about to show how money started off then everyone starts to think of where the tale of money started in our using of it. We think in terms of what traders offered other traders and then those traders started using a system where some easier form of currency had to be introduced to make purchasing and selling commodities easier. We think of honouring a genius unknown to us today that had the inspired vision by which to introduce tokens the people could put a value on and use it as bargaining devises to ease the burden of trading and thinking this is pure nonsense. That is what the Mammonites want every one to think but the truth of the matter goes back much further than that.

I wish to begin this by beginning to explain the biggest swindle man created. It is called money and in truth there was never anything such as money. Money does not exist. To have money one must steal from the earth and then swap this for merchandise others lawfully produced. Only by taking from the earth that which either belongs to no one or to everyone alike and illegally claiming ownership of what the earth produces can you give money any credence and value. The idea that money can form a basis as a visible valued factor is a hoax. Money is only taking what the earth holds and then trading that for what others have. However to put say coal into the trading system can only bring currency once and then the theft is converted to money. There can be no repeat of creating money because the theft happened once and the purchase it underwrote happened whereby the gaining of money is suspended into vapour mist. The coal served a purpose by coming into a system where it never had a place and it loses the place immediately by finding legal ownership afterwards. Say it was traded for a cow the cow carries the currency forward because the meaning of the coal is lost as a bargaining chip that others want or need. Then as the trader falsifies prices by raising the value of the trading commodity of money could money find further value and the truth is the process is lowering the value of the goods purchased or sold on. There is no honest manner that can make money more expensive except making goods less valued. It is more robbery by enlarging the role of the thief. But that is to the detriment of lawful commodities owned by honest people. The liars and the no-good-cheats are those praying to the money god Mammon and are named Mammonites after the god that they live for and the god who they uphold. Some call him Satan, others call him the anti-Christ or the Devil and in the Christian religion the Messiah was crucified because he fought Mammon by throwing out the offers that was for sale, turned over the tables of the money launderers and threw the money into the street as He whipped the merchants. Go on read the Bible and you will see He was crucified because He threw out money from the Synagogue and after that the Jews took Him to court. I guess every religion knows this

Mammon by another name the assimilation is representing evil. Yet I try to live without money because I am forced to and there is none tougher task.

I'll give you a precise example whereby you can obtain an insight into the mindset of the Mammonite and the way the Mammonite thinks and functions. I am a South African and being white being poor means having another definition about poverty than some of my black countrymen. I am dirt poor, as poor as a white person could ever get anywhere on earth but I have always got a meal ready to eat at mealtimes and a roof over my head because I have an educated wife that provides for our basic needs. I must say this at this point just to clarify my lifelong marriage. The only clever thing I ever did was to marry a clever wife and for as long as I live I never had to be clever afterwards because she takes care of all the cleverness I need. With her providing the cleverness, as she is the clever one I can philosophy as I do in this book but that is philosophy and not being clever. My philosophy up to now never brought in one cent because; well no one pays for information any more. It is freely obtainable from the Internet. With me also on top of everything being poor in health and one heartbeat away from the grave we have no room for luxury but in our daily needs we are provided for. In the black community in South Africa there are those living in squatter camps that has little or nothing to eat. These are mainly blacks from Zambia, Zimbabwe and Mozambique because the South African blacks living in desperations are entitled to food relieve and a tiny cash supplement giving the needy some income and their desperation also carries a different definition to what the blacks from Zambia, Zimbabwe and Mozambique experience. In order to bring relief and help to the desperate coming from outside our borders some Hyper supermarkets supply large containers in which people can place food they purchase extra and then donate that extra to the needy. These kind hearted Supermarket management then promise to take it upon them to deliver the donated food to the squalor camps and bring relief to the hungry living in desperation. So why are they Mammonites because after all they form a channel through which food are distributed to those with food shortages. You see those I mention as the most generous Supermarket management have these enormous bins in which the bleeding hearts can put the food but the bleeding hearts purchase the food they donate at full shop prices. There is no line or special till that you can run the donated food through where the till will give discount of any sorts towards helping the public to donate an even larger amount of food. If you buy even in the event of buying for the needy they see to it that you pay full price and in cash! Those being the most generous Supermarket management ask full price for whatever the bleeding hearts purchase and then they make a large profit on the sympathy of the bleeding hearts as well as the desperation of the hungry and needy. To them being the most generous Supermarket management the channel to provide for the needy is only an outlet to encourage the bleeding hearts to buy more and therefore push the most generous Supermarket management's turnover profits higher. The Mammonite takes advantage of every situation to enrich the bank coffers of the miserable cheats and line their pockets with larger sums of money. They urge others to purchase and donate towards the needy so that they can sell more and make more money from all other person's needs and kindness. That describes the Mammonite. How did we allow these bloody bloodthirsty bloodsuckers get their paws into our society?

When economy as a trading device started many years ago there were those that was willing to work and those not so willing to work and today these two groups are still with us dividing our society into two groups. The not so willing to work was thieves and low life and those willing to work were those honourable men around which culture and the future pivoted. The two groups are still amongst us with one being the working class and the others being the Mammonites, those holding money as a religion and praying on the rest to steal and to gain from what others work hard for. The grouping has no connection to religion because there are Jewish Mammonites as there are Christian Mammonites as there is Hindi Mammonites as there are Islam Mammonites and those religions I left out also carry the same ratio of robbers, as there are honest workers amongst them. Mammonites create a religion called serving the god called money and this is the true religion they serve. It is not a reference to a group of people or a renaming of a certain religious conversion but to show whom amongst us are the parasites.

When going on a discovery to find the root of all that is evil by representing dishonesty in all of man or women I have to begin where civilization began and that was (I supposes) just after language became a tool of progress. Everyone was a hunter and a gatherer and killed to live. Then the more intellectual saw that by not killing but herding animals they could prosper and find a more easy life. Farming did mean being busy most of the day and guarding the heard at night but better security for the goodwill of their children made it a worthwhile option. Some hunters saw an advantage in not killing everything but started caring for some animals they caught and looked after. They found a way to protect the animals by guiding them to better pasture where in the minds of the animals they were better off when being protected from scavenging wolves. Others saw fit to put specific seeds in the earth and harvest the crop at the end of the summer. Some saw this too be more cumbersome than hunting and felt they could become parasites by dishonesty, which is better known today, as a merchant's cunning trade ability. It still is robbery and it is dishonesty. The merchants were too god damn lazy to work and swindled the honest person by cheating.

This also meant by farming with animals or crop they were attached to the needs of their animals and could not roam as freely as those that chose to hunt but hunger was less prevailing when more effort was applied to look after the animals that they herded. There was not much free time on hand but there was job satisfaction in the price they paid. The rise in numbers of the animals brought more than their needs asked for and a surplus arrived which stood them and their children well as they could prosper.

However those not willing to work had a lot of free time to gather what the earth provided and could harvest what the Mammonites never planted. They could steel from nature and trade it off as property owned. Those not working could gather from the earth things that were not theirs to gather but they took it from the earth anyhow. They got hold of flints and metal and things the farmers needed and the farmers had animals the Mammonites needed to survive with in the harsh icy winters when hunting was hard. Someone had to have a plough, someone had to have an axe, someone had to have a knife, and someone had to have spear points. So, people had to have something manufactured. When everybody had animals, the one had a cow the other sheep or chickens and then maybe said the cow was worth a hundred chickens. So, they started trading. A cow was traded for ten sheep or a hundred chickens. Everybody was happy. It was a trade off. But no one realised anything. There was no growth. It was just trade and everything was resolved by swapping. Then someone at the very beginning discovered flints and thousands of decades later this became copper and then iron. Now there was a farmer who had ten sheep and he wanted an axe, which at the time was a wooden handle with a large stony flint tied to it by goat guts. So the farmer needed a flint to chop wood so that his wife could cook food. The brilliant clever or cold hearted criminal depending on who's point of view we find a description got a flint for free and brought this to the farmer that worked his arse off for one year to produce what he harvested albeit in animal form or in plant form. The farmer being honest thought that the criminal merchant also toiled for one year as the farmer did to produce a crop as a product and with that in mind had an idea that what the criminal offered had much value. It took the farmer one year to produce one cow and he traded this for one flint because in his honesty the farmer did not know any better. This enriched that lazy filth forming part of our society as the merchants. The Mammonites never had to work but only pray on the needs of the honest and then rob them blind because the working class were the honest and sincere group. This carried on while the dishonest Mammonites prospered to begin becoming the rich Barons and later Kings and forced to honest people to submit to the Mammonites' gangster enterprising. The Rich criminals became the powerful because the dishonest purchasing brought the imbalance that still prevails. Any Hypermarket buys from the farmer what it took the farmer one season to produce and then this the Mammonite sold to others for ten times what he pays for it while it takes him hours to sell from the point of purchasing whereas it took the farmer on year to raise the crop he sells for one tenth of what the criminal sells for. How do I know all this; I was the farmer and I sold to the merchants in this manner. This practise of cheating and plundering became commercial progress and all honest persons suffered equal at the dishonesty of the criminals that set up shop with enslaving the honest as their main intent.

Let me explain the difference between the mindset of the honest persons and the thinking of the Mammonite furthering his criminal intent. Where it began in the beginning when money became a tool of the oppressor to construct a basis for the rule of the false and dead god the farmer had to plant crop to find sufficient supply to cover his needs and the needs of his family. Before farming became a structural enterprise was to go into the wilderness and locate fruit and then harvest the fruit but this practise took up too much time to make the effort viable. Stealing from nature was no option that lead to survival and therefore harvesting the seed and preparing field in which to plant sufficient quantities was the only way to go. If the farmer wished to survive from the trade he chose he had to put in time and labour to harvest enough to allow his family to survive. But enough was also to harvest more than was needed for personal use because one can't live on bread alone. So he had to harvest the crop he chose as his "bread" to find some "butter" to add as spice. He had to harvest beyond personal need to trade with other farmers whatever they had as farming produce in order exchange what he had for what they produced to give his family and the family of all other farmers a balanced diet and in this manner everyone found a balanced diet by eating more than "bread" alone. Self-supply extended beyond personal comfort and went to providing for the need of your neighbour as well and in that trade became a manner of being civil. That good heartedness made way for the plunder of the Mammonite as it does in current times.

There were other basic needs the farmer had to employ as tools. The farmer used an ox to plough but while using the labour of the ox the farmer cared for the ox by providing fodder and water in order to tend to the ox because the good health of the ox was detrimental in supplying his family with essentials. The farmer could not train a new ox to plough every year because it takes years to train an ox to do what is intensive labour on the part of the animal. Because of this he had to put a retired ox out to pasture to use the old ox to teach the new ox how ploughing was done. It was a long-term investment and generations of oxen were needed to sustain the present and all had to come from the intelligent planning of the farmer.

Sure the Mammonite also required intelligence but that was being shrewd and cunning and always planning to steal rob and swindle. If the fisherman wanted to fish as the farmer did on a daily basis the fisherman had to cast a line and pull in a fish. This made the fisherman harvest what was in excess in the pond. But then the Mammonite-syndrome kicked in and the fisherman thought it would be better to go to the middle of the pond or lake and harvest more fish over there. Fishing from the side meant the fisherman had to supply some food at a certain spot as to lure the fish to that spot every day and those fish that didn't get caught got fed along the way. If there was more fish he could catch at any given point then it would be in the centre of the lake and going to the middle of the water brought more fish but the harvesting had to use a method that required money. Why is that you may ask. To get wood to build a boat the fisherman had to get a woodchopper to get him a tree down. If he chopped the wood by himself he could not have time to catch fish and his family will then starve from hunger. So he employed the services of a woodchopper. The woodchopper needed an axe to chop the tree to get the fisherman his wood with which the fisherman was going to build his raft with which he was going to fish in the middle of the pond. Therefore fish had to be traded to get a big enough flint to do the chopping of the tree and this brought about collaboration between trades because the fisherman had to catch enough fish to give the woodchopper to feed his family while he was chopping the tree and also the fisherman had to fish in sufficient quantities to give the Mammonite what the criminal wanted for what was not his to begin with.

The next question you may ask is what is the difference between the sleazy Mammonite taking from the earth flint that is not his to take and the woodchopper taking wood from the forest which by all account is not his to take? The money par comes about when he can take as much as he needs for personal use but not to swap it for other commodities because that is creating money and plundering nature. He will inevitably plunder nature because when he takes for trade, greed kicks in and lust for wealth makes that his consciences suffers and the penalty that must be paid for such lust and greed is the earth being demolished. The difference there is between what the woodchopper does and what the Mammonite does is no difference as they both plunder nature and take what is not theirs to take. This practise will engage the intervention of the criminal Mammonite since the woodchopper will find some Mammonite that will buy the wood for next to nothing as long as the woodchopper delivers it to his shop's door where the Mammonite then will sell the wood at an extravagant price because everyone perceives this criminal bent on enslaving others as being "clever" and "bright". The Mammonite rapes honesty and then feels confronted if anyone calls him by the trade of his character in being a criminal. The trees are not theirs to take an when they "have" to take it then don't turn it into money by giving the tree a purchasing power. The "taking of what is not theirs" and "turning that into purchasing commodities" is what creates the money aspect and the money kills the tree by having the trees become merchandise. Britain once was one big forest with trees so abundant there had to be woodlands no person could enter. It has been desecrated by human enterprise to a point where it now is largely savannah. Everyone plundered forests at will and nature could not replenish in the numbers humans stole and now wood is shipped from South America and Africa where these two continents are turned into desert, and yet the Mammonites are with us still, ruling our politicians that we supposedly chose by democracy to rule us but instead they rune us. If you think how Darius the Great ruler of Asia in antiquity and his sun Xerxes I built on two separate occasions during the two Persian invasion of Greece a sea crossing formed by wooden rafts the wood they plundered changes all of Asia from a forest to a desert. The wood used in times of antiquity seems to have changed mighty forests into what we now have and that is never ending deserts.

As time progressed many diversions of farming became practise. One such an aspect may serve as a very good example to show the influence commercial intent had on civil development as time went on. The farmers planted vine to grow grapes. The grapes were exchanged for whatever the vine farmers needed to live until the next crop was ready for harvest. The vine farmers liquefied some of the grapes to produce wines and the wine then was exchanged throughout the year for commodities that would sustain their needs and livelihood. Then came the Mammonites and they saw what they call business opportunity as they became wine merchants. They bought wine from farmers and sold the wine on at say only ten but most probably hundred times the price they paid the farmer. The farmer had produce that was bought. The wine merchant exchanged the produce for money by elevating the worth of the commodity and by this corrupt enterprise created money that in principle is and was absent when the farmer sold to the merchant. The Merchant added nothing in terms of labour or any inset in multiplying existing quantities but elevated the prices to create money, a means and a commodity that does not really exist.

If the fisherman or the woodchopper had to produce a tree and then to produce from this tree a boat by which they could go fishing the fisherman and the woodchopper would starve of hunger because it takes from thirty to three hundred years to grow a tree that can be used to build a raft that can be used to increase the income of the fisherman. So they do what any corrupt-minded, un-scruped Mammonite do and create money by stealing from the forest wood that belonged to no one or everyone alike. They took from the forest, devastated what was in the forest for all in time to come, traded nature for money that does not exist and messed up everything just to produce money once and with money everything else disappeared into nothing. This way of civilising civilisation brought progress and progress decimated civilisation by creating a need for merchandise that contributed to more plunder because creating money created a required sustaining of plunder, of which mining is a result practised today as it never was before and now in this text serve as another very good example to show what they take what is not theirs.

Progress brought about the mining of metals. How can this be dishonest? Then there was a farmer who wanted a saw and there was a farmer who wanted a plough. And there was someone who had iron. At first it was flint but someone got hold of copper that was then replaced by iron to become the marker for the Iron Age. This applied in the flint or Stone Age, went on to produce the bronze in the Bronze Age and gave a name to what then became the Iron Age. I fill this in to show as humans progressed all humans carried the burden of what the Mammonite-mentality taxed honesty through all time that formed civilisation. But the iron was not his. He sold what he took from within the earth by recovering the currency he used as trading merchandise by recovering iron from the earth. That was how money started. The currency started with things people took from the earth that belonged to the earth and that person's claimed what was not theirs to claim. The Mammonite decided that the flint from which one could make an axe is worth a cow. The farmer that had no axe and wanted the axe was then one cow poorer and this dishonesty was securing the value of money. He swapped the axe for a cow and the merchant gave a sword to another farmer in exchange for five hundred chickens. The cow now was worth five hundred chickens if you wanted a sword delivered by the Mammonite but it was worth five cows if you wanted an axe delivered by the Mammonite because a cow use to be worth a hundred chickens until the Mammonite got the help of the King and taxed all purchasing of essential commodities and with the help of the King and his clerks the prices moved up faster than the farmers produce or could trade. In the hands of the farmer one cow was ten sheep that was hundred chickens but when the merchant entered the trade market five cows was hundred sheep and that in turn was five thousand chickens when bought at the merchant and which was very legal and not only was it legal but it was no longer legal to buy from the chicken farmer or the sheep farmer or the cow farmer but it had to be purchased from the merchant just because the merchant was in cahoots with the King and the highway bandits alike. So the highwaymen became tax collectors while the Merchants became advisers for the Church because they too came in on the act. The Church and the priesthood were trading in soles and in sins. There were soles to buy and sell and clemency the people had to purchase so that the Devil would lose the fight for their eternal soles.

Farmers became venerable because of criminal gangs becoming the King's tax collectors and while also collecting taxes they plundered the rest for personal profit. In order to protect their interests the farmers needed swords. So the farmer gave in to the weapon merchant's hugely inflated prices to obtain a sword. This forced the other farmers to give the merchants a top price in sheep or goats because if the farmer didn't have a sword and an axe, jobless marauders would kill the all the farmer without hesitation because the merchants paid the bandits to raid the farmers just because it was good for business. Today this is called the insurance market. Now in view of overwhelming danger the farmer decided a sword is worth a much as twenty cows because the farmer needed protection and that service was part of the global trading practise. But everybody knows that a cow is worth ten sheep and a cow is worth a hundred chickens. By starting to promote violence against those unarmed farmers not willing to pay exuberant prices the hypocrisy of the traders became a battle to arm the bandits for free to get the farmers to pay for weapons and that the farmers had to do since the farmers were desperate for means of self protection. So a sword became five hundred chickens and a sword became 100 sheep and it could fetch ten cows at your local merchant. If one farmer is willing to pay up then the rest must follow the inflating trend of prices and shut up because plunder and theft are the way the economist work. They employ the Kings guard as much as they empty the jails ands get the jobless scum with small minds to commit murder and supply them with booty. They take what is not theirs to take, albeit stolen goods, looted goods, King's ransom and miners plunder it makes no difference because the lot represented money. This had to be money because if it were legitimate goods it would have had a trading value where one person that has something he harvested gives someone else something in return for something the other person had to swap. We see this with the trading of oil, food and electricity and the mining of gold even to this day.

Then there the notably honourable economist had to go dig in the mountain and retrieve more iron core, which they melted and produced metal and started to provide everybody with iron and start a civil war. This was labour intensive and we know the lot that did not go farming was too lazy to farm and they then will also be too lazy to mine! Now came the point where the economist had to perform in much brilliant wisdom. Right now we have the economist ruling the world by paying the politicians to help with the plunder of the commoners but back then it was only the King they had to approach with lots of money. But lazy as they are while they think it is intellectual the economist-Mammonite got others to do the digging and do the labour so that this Mammonite can feed on others labouring. This was a bloody good enterprise. Those the King did not kill in some war that made the King feel as if he was equal to God could die as young as if in the King's army by mining for the Mammonites that was serving their god they named money and was well fed on sheep-shit and cow dung because the miners had to repay the Mammonites for such kindness and sacrifice as performing job creation. The Mammonite gave the gold-digging labourer sheep-shit to eat and expected gold billion in return. The sheep-shit the Mammonite got from the farmer for free so that the labourer came living and working very cheap. To pay for the minor's labour the economist made the miner purchase the sheep-shit at a company store and made him pay through the roof for what the farmer threw away. This is admirably clever and cunning financing practise.

The cow farmer had his sword, the sheep farmer had his axe, the chicken farmer had his metal piece that he wanted and the metal piece did not run away. The economist with his iron had a problem. Trade went slow because the merchandise did not move and the need for purchasing power did not grow as rapid as the merchants grew in numbers. So as all economists do, he went terminal. The people who did not have cows or sheep or chickens, the economist told them that they could have the metal piece they wanted if they brought him 2 cows, 20 sheep or 200 chickens. Then those that had nothing could have the sword. The guy had no animals. So he went and robbed the daylight out of the farmers. He brought this guy his 20 sheep or 2 cows and he paid for the sword. So money was created in terms of robbing those that have and sell to those that want to buy but do not want to pay. Now, suddenly the sword was worth 2 cows, 20 sheep or 200 chickens as long as the one that had nothing had animals with which to purchase arms. That was the first time inflation or growth was created. He who had nothing stole from those who had something to give to those with criminal minds also known as criminals or economists and the economy started growth. The economist had those who had nothing on his side, supplying them with weapons to kill those who owned something to bring to him who had the metal factory and when this happened we had the first bankers. In all of this the King was paid to guarantee protection in order to maintain free trade amongst those that had nothing but always had something to sell. Today we call them shop owners. The bankers have animals deposited at their doors for whatever merchandise they had available and whatever they had was illicitly obtained from those that had nothing but always had something to sell. Now the banker or economist had another group of people who did not want to rob the farmer, kill the chicken farmer but had to eat. So the economist gave those people labour. The economist had to get something in return. He put them in the mine to work. And then at the end of the week he gave them a chicken stuffed with sheep-shit as a bargain just to keep them alive to retrieve iron core for another week.

He gave them just enough so that their children will starve and go hungry but the workers will stay alive. Seeing their hungry children starve was motivation to work harder to enrich the Mammonite even more and that spurred the Mammonite of to let the children become ever more desperate for food and that spurred the workers on to become more loyal and work harder for less benefits. In this money surplus grew. So, the biggest criminals on earth are the economists. They kept slaves to feed their economy and had the criminals who kept the economy alive and they had the worker which they gave work by job creating to retrieve the iron core, which was not the property of the economist or Mammonite to begin with. The more iron miners retrieved the more money there is because it becomes money and the faster the criminals steal from the working people and then the working people had to purchase again to replace what they wish to protect but what they will lose in any case in the next robbery the criminals do on behalf of the economists. To be able to keep the economy flowing the economist had to keep on retrieving from the earth with mining to create money. This they did by stealing from the earth, that which wasn't theirs.

And they swapped it for farmers that farmed the cattle and there was a normal flow of increase in cattle. Only when you retrieve that which is not yours, you can have money. Otherwise you have equilibrium. The one will trade his cow for ten sheep, tired of eating cow meat. The other one will trade a pig for hundred chickens, because he is tired of eating pig. But there is a normal swap where nobody loses much, but nobody gains much either. The one that can go into the bush and shoot the pig and came back and traded it for a chicken, the farmer then did not eat his own chicken or boar. The price remain the same – therefore equilibrium. But the hunter living in the woods needed iron. He needed a bow and arrow to shoot the pig. So, he started trading for little. The miner got for nothing because what he used as currency he was stealing from the earth. And he was using people to mine for him...because he was too god damn lazy to work. Then he decided on money. Money is the root of all that is evil.

Now there was a prince. The kind ruled over all these peasants. But the king saw another king having a kingdom of his own. And he had thoughts to fight the kind and gain his property. Then he decided to buy swords. But to purchase these swords he had to do what the merchant wanted. The prince thought it wise to listen to the Brainy that had insight into forces that pulled and pushed weather conditions, could calculate and command lightning and could order floods just by studying the stars. It would be much better for the Prince to have this lot on his side than to care for the beneficial conditions of the hard working farmers that was already over burdened with taxes and contributions to the Crown and every war effort the Prince could dream up. If the farmers started to get grumpy the Prince sent his army in and pacify them into silence and obscurity with sword wheeling and knives cutting flesh to the bone.

In this it is not hard to see who the Prince would favour to carry the banner on his side. These Soothsayers and Alchemists could order a fire by instigating a lightning flash and set the thrown on fire while the common farmers did what they were told and did it without quarrel. Could these wise wizards do what they said they can...no but it was better not to take a chance. So the Prince took from the workers to feed the Scientists that could see forces pulling and forces pushing and Alchemists that could turn lead into gold and paid the farmers money while taking merchandise in the place of the money.

Doe any of this ring a bell still being part of our modern wise society. Bloody hell yes, it is our modern society. When Julius Caesar learned of gold mines in Europe he attacked the countries that had the gold, labelled the Kings and people of those countries half-witted barbarians that carried the torch of Satan and was enemy number one of the state of Rome. Therefore to create enemies of the state that could offer gold as ransom and prevent the Romans of paying taxes while the Romans could party in the Arena made Julius Caesar a very memorable man. In fact it made him so famous we know about him as if he was a modern hero. We know him so well because he killed people by the hundred of thousands. We think of him in great respect because he robbed and plundered those that had mines and had money and while he ransacked their villages and raped their woman while slaughtering the men he goes down as one of the greatest leaders in history.

While being one of the most blatant criminals, one of the most outrageous robbers, one of the biggest plunderers of all time we revere Julius Caesar and all others just like him with awe and admirations and we stand breathless thinking about their greatness. They are remembered in fondness because they could rule and create riches. That is a bloody lie. They stripped mimes and minerals that belonged to other nations and murdered them to get their hands on the loot. If they did not plunder in excess we would never have heard of them, and still we remember their greatness in fondness. There is no such a thing as money and money can only be if it is gains as a result of being stolen and murdered for and received by applying the most brutal crime. Luckily that was then and the last such hooligan was napoleon that went through Europe in the name of peace and plundered everything he saw in the remembrance of bringing in civilisation to those that had none.

Remember George Bush senior and the nineties and his election promise of not raising taxes when he did just that just after the elections. Remember the words " read my lips" and he still backtracked on his promises and got the voters raging mad? When George Bush saw he was in a pickle he turned a friend and a partner in war into a household name standing for everything evil Saddam Hussein warred the Iranians with weapons the Americans supplied and bombed the holy shit out of the Persians on behalf of Israel and America. Then just as Bush realised he made a mistake that was going to cost him his second election he found a foe worthy of war and to top the lot this country had oil in huge supply. Then

Bush bombed the shit out of Iraq and lots of reasons showed they had no necessity for a reason while they got the war machine going that stimulated the economy.

I can go on and explain how every leader is a war criminal and how they went in to kill Muammar Gaddafi and by bombing cities to bits destroy Libya just to get into a position where the lot had to be paid to "rebuild" the ruined cities and get money for the rubble they brought about in the country that they shot to pieces. The worst crime of all is that money does not exist. All the murder and the plunder is to create something that never was. There is no such a thing as money. If you shout "Gold!" so, what about "Gold!"? I ask you where is the hundreds of tons of gold that the Romans went on to kill almost the entire population of Europe? What happened to all the gold that filled the coffers of empires in antiquity? Where has to gold gone to that Alexander the great retrieved after he defeated just about the known Universe during his lifetime? What happened to the hundreds of tons of gold that Genghis Kahn sacked his world he lived in? Where is the gold that the Spaniards brought back from the New World and that was not that long ago? Spain in its present state is economically almost a failing state and yet it brought home riches in the amount of silver and gold by the shiploads. So what happened to that gold because it is no longer in circulation? We don't see the billions that came to Spain or the Netherlands and these countries floated on wealth not that long ago!

Moreover where is all the gold hat Britain stole hen they murdered a third of my people in concentration camps just over a hundred years ago? Where is the diamonds they sacked from my country when they raped and plundered my people in the very same manner as the Romans did and the Mongols did and the Huns did and Napoleon did not to mention all those I did not mention? Where is the wealth that was created by theft through out the ages? It is gold and it is silver, which is a metal that can't be eaten and can't disappear into mist.

I wish to show you the reader what money is and you can try it in your life. When you have something tangible and solid and you wish to sell it that would mean you wish to exchange it fore currency. In this instance I am not putting you in the shoes of a trader or a merchant because I have explained that part. I am talking about a situation that will come up in the Every-Joes life as he or she goes through life. I will take my selling of my farm as an example. I had to sell my farm because at that stage I had a better chance of dying from a heart condition that I had surviving the heart attack that I was waiting for knowing it had to come. I could not leave my family in the position in which I was knowing they ahd no chance after my death to get anything of value from a sale as part of my estate with tax and all that being part of the reapers harvest.

The day I sold my farm I told the person purchasing it he had a deal that would never repeat again. The moment I opened my "hand" to accept his money I lost about a third of what I had before the transaction concluded. The moment I accept money and I wish immediately the next moment to return the money for what I had before the transaction the buyer will never accept the money that he paid for the commodity. The moment you take the money you lose about a third of what you had before the transaction concluded. If you wish to buy back what you had you will have to pay about a third more to make the new owner willing to part with what he just purchased. You may put the money down on a new house or a better farm or whatever but then you only pay a certain part of what you had. You then will have to work for many years to gain what you have lost and you may in the end have more than what you had but you will work for decades the be better off than what you were before the transaction ended with the exchange of money. The second you accept money for anything you have of value you lost money that very second, that is if you did not swindle and falsify the condition of what you sell to get the person to pay more for what you have. But when you do that you get on the side where the devil or Satan is and that is the side of the person without scruples. Then you become a Mammonite.

In the present time things are looking up financially all over the world because humans found a way to plunder the earth from energy it stored millions of years ago and had the sun in storage by the way of carbon fossil plants that went to become coal and tar and oil. It is this plunder of fossil wealth that puts the world in such a fine place in which it now seems to be. They take minerals and metals from the earth as if there is no end to the volume in storage on earth and their greed makes them mine as if it will last another ten thousand years…and that is to gain money that disappear as soon as it is mentioned. Yet things are as good as it never was before this time and things can only get better. Man is so clever there si no stop to the ability man may have as long as man does not run out of money. We are a clever lot that can discover anything and create what we can only dream about.

…And there is the problem we have with money. Man can create nothing and that is all. Newton created nothing and Newtonians put it into outer space in such quantities if is overflowing forming the expanding Universe. Money is the only true thing that man created and that is the proof that man can create nothing, not even intelligence! Man thinks of his position as sublimely intelligent and crafty, which is everything that man is not. The only thing that puts us in a better position than our ancestors were in the past is we found a way to harvest energy that was stored from the time life started. It is the sun that sends us heat and it was locked into life and the life turned to carbon and now we burn the carbon to get heat that was heat that came from the sun over the past few millions of years. If it took the earth that long to sore this energy and we waste it in say one hundred and fifty years we are heading for lots of problems which is surviving as humans in time to come. Everything is plundered so that a few fools can feel how it feels to be so incredibly rich as no one ever felt before and then they die and all that wealth is lost. Just like the gold that was bought into money in the past this money exchanged for oil and coal and gas will vanish as soon as it comes into circulation and it will stand to benefit nobody in the very near future because it is money and money is the creation of man and man can only create nothing!

In the society we now have I can only see idiots and more idiots. Look who are our heroes! Pop idols that can sing if you call it singing. We have movie stars that are gone in sixty seconds as soon as they displease the money moguls that promote movies and financially harvest money. They are created and they are the ones everyone looks up to. They are adored and they are made to please as long as they please. They fill the printing press because the printing press publishes what the idiots with no minds wish to read. You open a magazine and you see some dumb-eyed blond that I accept must be a movie star or a singer or some nut living a glamorous life and she parades in a swanky dress that she will never where again. So the money paid for the dress is gone and her celebrity is gone as soon as the film becomes yesterday's news. However this is not the lot that carries our society into the future. They are not forming the absolute pillars of our community...no we have bigger fools with even much less brains carrying that torch and filling all the promise the future of man might hold.

If you wish to be the pillar of the human intellect you must be able to kick a ball...or hit a ball...or throw a ball... or do something with a ball! On those that carry fame rests the ability to do whatever is funny while doing it with a ball or something that can replace a ball such as a pug or whatever. If you can kick a ball you are paid in tens if not hundreds of millions and woman throw their virginity away after you and men sing your praise and families carry your name or the name on their chests or the brand logo of your team on their shirts just to show all other persons they wish to be associated with someone with your capabilities. What a lot of crap and that hold the intelligence of the human race in ransom. That is the mindset that will carry intellect into the future. God will help me by getting me mortal and dead because I know by that time my age if not my poor health will see to it that I am dead and I will not be part of what is coming in one hundred years from now. Boy is there sorrow coming when this what everyone believes will never end is going to collapse into a mountain of stinking hog shit. If you can do something that no other can do with a ball you are revered, as a super human being although you are so stupid you can't talk properly or write you name in a sentence or even understand the most basic mathematical equations. Still if you can do special things with a ball that others can't copy you are thought of as holding company with Jesus Christ amongst other big names and you can take the place of God Almighty whenever you choose. This lot with that has a substandard mentality and that has no understanding of facts and are blessed with such meagre wits are the ones on which the society has to build to secure the next and the flowing and the generation after that. By that time we are t of oil and out of coal and out of food and out of intellect. This idiot that thinks a soccer or rugby player is special is the salt of the earth and that intellect and understanding of expectations will secure intelligence, which will maintain civil order in time to come!

I know you are asking what is all this money got to do with science and this is what it holds connection to.

I can much more believe in a ghost than what I can believe in money. At least I can create a believable ghost in my imagination but in my imagination I will never create money. Money is a reality that can never be while a ghost might be a reality if I wish to make a ghost a reality. We all, and that include me live for money and especially me that has no money. I think of and I think about money night and day because my only fear I have is to be without money because I am constantly out of money all of the time. If I want a cold drink I have to ask my wife. That is how desperate I am for money and yet I hate money because I know money is the root of all evil and money is the force we call the devil or Satan. Money is the Anti Christ everyone is waiting for. Still try and go without money and you will find how little pleasure life does offer when you have nothing to pay with for any pleasure you want. We all pray for money as much as we fight for money and we are all slaves of money for what money represents while money is one thing that has no value and is completely fabricated by swindlers cheating the human population into direct slavery. For every job Mammonites create another slave is put to the plough to enrich the greed-grabbing slave drivers that want everyone to see them as being these pleasing-everybody, well-to-do and so kind-hearted money-monsters. Money and what is connected to it is a phantom some make a reality to enslave those that have no money and get us to work for those that has money while there has never been money. From forming a society we were engaged in believing something that can never be.

Let's follow the dots and find how this connect with the dots and see where this connects to science and in particular basic physics. We live for and we live by a commodity that does not exist and never had a value and yet in thousands of years millions if not billions of lives went lost to gain money and the power coming from money. The First World War gave up twenty million lives for a fight about money and who ever won the final battle got land and riches from the loser. It was about the one side getting rich to the loss of the other. This is what all wars are about. The Second World War saw fifty thousand lives go lost and if we as humans are prepared to lose our lives for something that does not exist, how the better will we believe in science that does not exist. We except what we are told just because we are told and if we question that which we are told we are true.

In the next pages the tone is getting more complex but it is not difficult because it is I that introduce these ideas and I am just another ordinary fellow with ordinary abilities. What does make it complex is that the reading requires much concentration because it is new ideas that you are going to engage with for the first time ever. Because it is new it requires much more contemplating than what is normally required when reading things that you have heard a million times before. We are brainwashed to except that the only undoubted truth we can ever come across is science. That is where the fable starts because in ninety nine percent of the cases what you read as science is the opinion of someone that considers his opinion to be factual because that person sees himself fillign the centre of the Universe. It is just another opinion and most times those that call then scientists inform you about their opinion without showing their case studies so that you can use your intellect to form an opnion on the matter.

If science cannot prove God's existence, it is not God that does not exist, but it is science failing and therefore it is then that specific view about science that should be re-examined since it is the view on science that is proving as being incorrect. This fact is what the so very brilliant and intellectually mindful Newtonian atheist should remember when they fail in their science altogether. That their science fails altogether and that failing it does in all its splendour, is facts I am delighted to prove! The fact is Newton's views were never tested and that the Newtonian views on science were never challenged before and because of that Newton principles never withstood diligent scrutiny before.

When **Sir Isaac Newton** is investigated even in the flimsiest of manners, well accepted facts seems to become very suspect, to say the least. This becomes evident when concluding all the facts this book presents. Now, in this book, for the first time Newton is tested and such testing is the proof you gain by reading that which I uncover. What I bring into the open is unseen facts, which I present you with as I take you on a tour through an avenue of facts I introduce in this work. The lack there is in sensibility concerning **Sir Isaac Newton's** principles this book proves. The theories of **Sir Isaac Newton** require proof, which was never given while God never needs proof and that is what science constantly seeks. When science perpetually ignored my concerned calling on and ignored my calling on them because (I suppose) they were finding my concerns wanting, in my final letter to them I promised them never to contact them personally again by any and by all means. I also promised them a fight. This is the fight I promised. I was not worth noticing so I was ignored. I now am calling on the public, as I am ignoring their reputations. I am showing the public just how extremely bright the Newtonian inspired super-thinkers are!

Scientists portray the image that they know all there is to know about everything man might ever know and nobody can ever know more than what they currently know. They tell you what to eat, what to drink, what to think and how to live because they have this image that mentally they are on par with God. They are the utmost superiors on all levels of what forms creation. This attitude applies to Rocket scientists as much as it applies to medical doctors as much as it applies to lawmen. This comes through in all departments albeit language, art, science, law or whatever you may have. If you smoke cigarettes then you will die young because the medical profession found that it is harmful to you while a hundred years ago people smoked ten times what they smoke today and they lived just as long as they live today. But the medical profession took it on them to act as God and force-feed everyone to do what they say or die. The truth is the oxygen you can't live without is burning you to death by aging you and without that you die. If you don't breath oxygen you die and if you breathe oxygen you die because you are born to die! In the end oxygen kills more people than any bad habit or disease because it slowly kills everything with life. This acting like God envelops all forms of science. What is the truth? Do we hear the entire truth? The truth is that science only reveals some portion of what they know and ignores what is there that they know they don't know. They only reveal what suits their position and never divulge what they know but what does not compliment their view. When science confronts religion they have the opinion that what is in science is everything there is and there can never be more than what science knows or what science wishes to reveal. Science now knows everything knowledgeable and whatever will be known they know.

If some scientists are of the opinion that we will fry in boiling water in the next century then it is the Biblical truth because science holds the opinion. Science knows best! Today we laugh at medical practising of a century or two ago and in another century we know the future generations will laugh their heads off when listening to what the informed opinions are of the professionals today. Science has forever veneered their status with this blanket of "they know all". This makes a mockery of the truth because science has no clue why man die or why man age and yet they promise eternal life within the next few decades to come. Ask scientists what is life and they will have an "informed opinion" because they know everything there is.

Science keep up this front that they know everything there is to know while even reading my books prove how little they know about science. They withhold every aspect in science that they do not know and only elaborate in detail that which they think they know about and what we presume they know. They are of opinion that there can be no other way that creation started but according to their science. If the Bible describes how events unfold it then are incorrect because science knows everything. In **The Veracity of Gravity** I show **scientifically by using science** how creation started precisely as the Bible says word for word but then I also show how little science currently knows about science. The book not in print yet is **An open letter Addressing Gravity's Formula**, which is far more elaborate on the matter of how creation started where I show how science proves the Bible correct. How shocking this might be it is just as true.

Science can never take the blame for not knowing. Never is a suggestion put forward that it might be science that holds the shortfall and it is because of science not being adequate that science cannot match the Bible. I can prove how the start came about because I decoded gravity and I did that by finding an explanation about the four cosmic principles. By deciphering the Roche limit, the Lagrangian points, the Titius Bode law and the Coanda effect I am able to show how the very first instant happened when the Universe started the very first point ever formed. These principles are in place and not the principles Newton fabricated... That this book shows. It shows that the cosmos uses other principles than what the Newtonian science promotes. What science says nature uses is not in place or does not hold evidence while what nature does use science deny by just never pressing the issue. I show what is in place and I show why it is in place but first I have to reject what science says is in place because it is not in place.

I show how Newtonians fabricate Newton's ideas about gravity. This is ongoing since the end of the dark ages and Newton. There is no mass that can pull. Most people reading this and who are schooled in physics never heard of the Roche limit, the Lagrangian points, the Titius Bode law and the Coanda effect and these principles are what builds the Universe while I am going to show that there is no factor such as mass. While it serves their purpose notwithstanding never finding evidence to the fact, still science uses *only* and *exclusively* Newton's idea of mass while the principles in place the Roche limit, the Lagrangian points, the Titius Bode law and the Coanda effect are never ever mentioned. They sometimes put referring to these principles as law in brackets to deny the status that any of the above law have.

I am going to show you within the next few pages the silliness Newtonian principles hold. While I discuss the principles please see where I am incorrect or going wrong and convince yourself whom is wrong. This is because the Roche limit, the Lagrangian points, the Titius Bode law and the Coanda effect disputes Newton and science would rather discard what the Universe uses than to put a question mark behind the fabrication Newton put in place. Where everyone knows the fabricated information and hiding the reality, which is in place within the cosmos, and that is the conspiracy I show to all. Science stupidity ensures they don't understand the working principles that are in place and that was known for centuries in some cases as the Roche limit, the Lagrangian points, the Titius Bode law and the Coanda effect and therefore not knowing how the principles should be interpreted they hide the concept due to not want to be seen as the ignorant fools knowing the cosmos implements the principles as reality. Science hides their limitations and incompetence behind providing the public selective of information. Take for instance the edge of the Universe they talk so much about. There is no edge of the Universe because there is only an unlimited everlasting Universe out there. What the limits are that they see as the edge of the Universe is the limitation of their equipment that can't trace time back beyond what they see and that serves as their limit in understanding what the Universe offers and how the Universe unfolds.

Trying always be perceives as matching the likes of God science can't face the fact that they can't precede further into space by reading time than what their limitations and their equipment handicap them with, then they put their shortcomings onto the Universe having limits so that they can present the image of total superiority in contrast to limiting the Universe. If they do not understand the four principles in the cosmos and which the cosmos uses as building blocks how can they understand how the cosmos works?

Then because they are clueless about the information the Universe provide and because they use misinformed cosmic principles they're confusion gives the Universe an edge where it ends and a date when it started while they admit and we all know that the Universe is timeless and limitless in every aspect we encounter. One thing the Universe des not have is an end because I show where the cosmos holds infinity, the point that can never go smaller and where the cosmos holds eternity, the point that can never stop becoming bigger because it is endless.

When I say there is no such a thing as mass everyone goes quiet and I can hear they immediately question my mental stability. One may question the fact that there is a God or not and one may question which God is the true God and one may doubt any form of God existing but doubting mass or saying mass does not exist is utter irresponsible madness. Everyone knows there is mass. That is one thing no one doubts. The fact of science showing that mass pulls everything down is a real as being alive, or is it?

Well...I wish to bring to mind some of the facts that physics work with when academics as scientists only work with facts. Remember they are the ones boasting that if facts are not proven then it is fables and those very important academics don't waste time with fables because they only work with facts. The accuracy of their basis on which science rests is that mass is responsible for gravity by pulling. If you don't have mass you're not going to have gravity. Mass is equal to gravity and gravity applies only by mass. If mass is present then its by gravity or otherwise gravity is absent. If a body falls it is the mass that pulls the body to fall because the body receives gravity by ratio of mass and mass is that which produces gravity in relation to the mass available.

It is mass that drags you down because the mass is in charge of the gravity and the gravity finds the value from the mass available. Mass pushes you down by the gravity it forces onto you. But if mass drags you down then what lift you up in the air balloon? If mass gives the gravity to drag you onto the Earth then why would the hot air lift you up in a balloon? Is the hot air causing anti gravity or anti mass because gravity by lifting the air balloon and cargo. The hot air balloon is lifting the passenger and all that is in the bag plus the bag plus the balloon into the air. So what is then pushing the lot up if it is mass that drags you down. Has the air not got mass because then the air can't have gravity and then the air must escape into the blackness of outer space because by going up it shows a resilience of either mass or gravity. We have seen that it is mass that pulls everything onto the ground.

2   Einstein saw gravity from a window in a patent office

Years ago I was reading of a remark Einstein made about his realisation whiles being a patent clerk. Einstein realised that had Einstein fell from the window of the patent office Einstein would feel as if he was as weightless and as weightless as the chair and a pen falling alongside Einstein down the building would be. The only principle Einstein therefore could not accommodate in his theory on Relativity was Newton's gravitational pulling by the value of mass. Einstein saw this "feeling" as a psychological experience more than it was physics. This moment reading this gave me the breakthrough that I waited for and it (then) took me twenty-three years to make the breakthrough. It was not Einstein or the chair or the pen that fell but it was the space the three components occupied that descended.

Then I realised Einstein felt weightless because he was falling and part of falling was feeling what was happening to him. He was not pretending to fall whereby he then would feel as if…he was really falling and with that there is no as ifs. What he experienced came by means of what he was experiencing in as much as undergoing. If Einstein was experiencing weightless ness, it would be because he was weightless while falling. Weightlessness is not having mass and not having weight is feeling being without mass. If he felt being without weight he then was without mass and then mass has nothing to do with falling or the process of going towards the earth. Einstein would not imagine the weightless ness because Einstein was truly falling. He was at that moment truly weightless. He saw himself falling alongside (not faster as he should if he had weight or mass) than the chair and the pen that dropped at the same pace and at the same sped as he was falling. Einstein, the pen, and the chair had the same weight since they were all weighing the same because they were descending at the same rate. All three items would be equally weightless during the falling…that was what Galileo found because objects of different size and different mass travel equal while descending. This is what Galileo found and that is what Newtonian science can't reject but have to accept in spite of Newton contradicting this founded his pendulum theorem on. Newton contradicting this is because Newton claims things falls by applying mass. Galileo says all things fall equal whereby mass plays no role and Newton says it is only mass that plays a part. Reality TV shows that the bigger objects do not fall quicker than a smaller object and that can only be attributed to one fact; it can only be true if they weighed the same while falling.

From this one can deduct that gravity is motion or the intent to commit motion and mass is one the motion of gravity is frustrated by blocking the continuing of the motion. Gravity is motion of space and mass is the restricting of the motion of space. Having mass does not bring about gravity but it does restrict gravity's motion. Gravity produces mass but mass does not produce gravity. Mass is the restraining motion and gravity is material moving about. Mass only comes into the application when two objects filled with space moves into a position where both want to claim space the other occupy. In essence it still is the frustration of motion and the commitment to move once the blocking of space is relinquished.

I then after reading this realised that gravity is not mass orientated, but gravity is motion differentiation between objects. While falling, The object moves less or slower in the direction that the Earth rotates and will fall in the direction of the Earth centre until such a time as the movement of the object is in synchronising with the speed that the Earth spins or if not the object will and on the Earth surface at the edge of the Earth and that will bring about having mass. The gravity applies as speed that is putting time in relation to the distance travelled and distance travelled is space. While the object is in a process of falling, the motion confirms gravity, both by getting the object's distance or band in which the object travels in harmony with the Earth that conducts all the spinning taking place at that point. That will reduce the height in which the object spins until it lands on the Earth and then can't reduce such reducing of a travelling band any further. It has to do with specific density. If the specific density is increased by filling the object with helium we will find there arrives a point where the conducted speed is at a level that the Earth no longer will claim the body into having mass. When motion downward ends and the Earth disallow any further movement to secure a better specific density in relation to rotating movement, then mass sets in and becomes what is than point holding mass where the constraining of the object takes place to secure frustration of further movement and the Earth's motion annexes the object's freedom. While experiencing mass the motion is still there but now incarcerated by mass and locked onto the Earth by the rotation of the Earth and the superior or equal specific density of the Earth. By connecting to the Earth the motion that the object is experiencing is what nails to object to the Earth by the force of mass and the object is then experiencing mass and not falling further through the loss of downward movement and now only conducts with the Earth rotating side-on movement. In this the downward movement is not lost altogether but remains, as detectable movement is the form of having a tendency to move although the object in mass is applying by forcing the downward motion to stand still. While the object is in mass and seems to be as if it is resting the tendency to move downward remains applying but that tendency to continue to move downwards is the tendency he named mass. However mass then restricts motion and becomes motion tendency. While falling, gravity applies as equal motion to all objects relying to place all objects in relation to specific density and because of this motion counteracts any size, mass or weight by making everything able to fall equal in specific density. When falling, the object is either equal to what might be in the air according to allowed specific density, or has more than the specific minimum required density that is what is allowed to serve as the minimum required specific density and therefore will spiral down to the Earth. When the Earth restrains further downward motion of the object that comes as the result of finding an allocated position of motion according to the specific density of the falling object, this readjusting of allocated position is stopped from conducting further downward or readjusting movement and all such further movement of gravity is hindering in the form we call mass. The falling object remains individual and still tends to move while Earth individuality resists movement. Further movement is disallowed as other material fill space. While the bonding of the atoms forming the object will secure any further deforming the object will remain to be independent but it is this bonding that is the value of the specific density of the object applying. By securing a place on the Earth, the falling object will finally rest and from that motion resistance comes mass.

While falling, the object is experiencing gravity because the object is in gravity but when on the soil the object experience mass which is the restricting of gravity or motion of the space filled with material.

Moreover, I came to another conclusion of equal importance. When any person is standing on any place anywhere, while viewing the Universe, that person is filling the centre of the Universe. Let's get more personal. When you, the person that is reading this, are standing at night and are looking at the Universe you are seeing the Universe from the centre of the Universe. All the light, every single beam that ever left any destiny at any time acknowledges this fact. You are the most important person in the Universe because you are holding the most important position in the Universe. All the light that comes across all of space runs directly in a straight line towards you filling the centre of the Universe. Not excluding the effort of one photon, all light is heading to meet you where you are in that centre spot and not one photon will pass you by. Not one photon dare miss you because if they do they miss the effort that all light has to accomplish and that is to locate you as the person filling the centre of the Universe. If you find this funny, or laughable you are in for a shock because this is what gravity is and this principle dictates gravity. It is the most complex issue one can imagine and expanding on this thought takes thousands of pages. It forms the crux to all cosmic principles and embraces every successful and meaningfully theory ever used to explain the Universe. Without taking this aspect in to account, there is no valid explanation available to understand the cosmos. Al the light coming from wherever meets the point you fill in time and in space. For al the light travelling you hold the spot it was on route to.

Should you decide to shift your position to any other place in the Universe you will shift the centre of the Universe to that location as well. If you install a camera on Mars, the light is obliged to acknowledge your relocating the centre of the Universe at your will to reposition you're being that centre of the Universe. All the light that ever left its destination crossing the vast spaces of the Universe, excluding no particular light, travelled all the way just to find you filling the centre of the Universe, right where you are. By you're standing anywhere, you fill the centre of the Universe, and the entire Universe admits to that because all the light comes to meet you there. If you shift from the North Pole to the South Pole you will shift the centre of the Universe because all the light travelling throughout the Universe will find you where you then moved the centre of the Universe. The light left its destination billion years ago as it travelled through space at the speed of light anxious to acknowledge you're being in the very centre of the Universe. No photon will pass you by where you are in the centre of the Universe. No wonder every person born has the idea they were born to fill the centre of the Universe, which we do fill. The Universe is spinning around you or I, which is filling a centre where all motion is connected. That is the Coanda effect on the utter-most grandest scale imaginable; nevertheless it is only a manifestation of the Coanda effect. It implicates gravity as wide as can be…

Then I reviewed the Universe. If gravity is motion, what causes motion? What stops motion? That answer is in the Black Hole. If a star is about fusing atoms thereby growing, what happen when all the atoms fused into one all collective atom? What is the gravity if the star has one all-inclusive atom providing all the gravity that the star had when the star still had massive volumetric space? If all that space that once filled an entire giant star fused into one enormous gravity applying atom and that enormous force has been secures in the space that one atom holds, the atom would then show a force that would pull the surrounding Universe flat. Where does the gravity of the star end when all the atoms in the star became one giant atom? Gravity is smallest where space is least. Where space of an entire massive star is left in the size of one atom the gravity coming from that will pull the Universe flat at that point.

Coming to the conclusion about gravity being motion and mass being the restriction of motion was the easy part. What produced the motion and what prevented the restriction from overcoming the motion was the tough part. Figuring out why was everything on the move and where did the motion stop that was the part that took some figuring and some explaining. What made gravity move and why does gravity move…the answers are in the four phenomena never yet explained to satisfaction but now turns out to be the cradle of gravity.

Gravity is The Roche limit,
   Gravity is The Lagrangian system
      Gravity is The Titius Bode law
         Gravity is The Coanda affect

And gravity as the Roche limit forms the principle in producing the sound barrier. Read the book and find out why this is the case. I explain these in the next chapter or the chapter following the next chapter.

Newton's claims about the principles that he declared is responsible for guiding physics carries no validated proof and only after I realised that, was I able to start forming another line of thought on gravity. This had the purpose of confronting the corner stone of modern physics and at first I tried desperately to do just that. At first I was not confrontational towards Academics in physics and avoided any indication about disagreeing with Newton, although avoiding to show my disagreements was also totally impossible too but every time I approached academics with my new concept the academics always threw Newton at me . Facing Newton or facing defeat became a two-sided blade and I had to start to confront them by confronting Newton, with which I was in disagreement from the beginning. At first I was reluctant to voice any opinion about the matter of how far I was prepared to challenge Newton because Newton was and is an icon. But slowly it dawned on me that if I had any serious plans to introduce my ideas I had to dispute Newton's gravity principles and do it head. When the slight confrontation did not bring results I finally decided to go all the way and show the inconsistencies that were prevailing in Newtonian science. That worked neither and it brought me the same results as before whereby I decided to go public and straight to John and Jane Dow avoid arrogance academics have with only one motto they serve and that is their autocracy and in particular their megalomania especially to my case as well as me in person. I wrote them (nine in total) letters in which I warned them that I was going public to show the extent of their dishonesty in their Newtonian's approach and lacking of substance and proof their physics has. The lack of honesty and furthermore the absolute dishonest on their part is there whether I avoid it or attack it; the inconsistencies are part of forming the basis for modern accepted science.

This process I now described is explained in a paragraph or less and it seems I got that far in a breath or two, but getting this far took me the best part of seven years to get to I tried my best not to attack them or Newton but left with the option to leave the project and lose thirty years of work and then fail after I concluded an answer on every aspect they never even thought of or take them on and dish out what they should have received years ago made me decide on the latter. After being avoided and taunted by their powerful positions and arrogance vested in their mentality they show in regard to their positions as well as the disregard they show in the mentality of others I slowly concluded that only and after I can get people forming the general public and the opinion of those that holds their disregard just as I do to see what they hide will I get a response from the Mater's of fraud. First I had to show the general public the true colours of the academics in physics and get every one to see how incorrect Newton is, and only then do I stand any chance to introduce my line of thought. I am so sure of the ideas that I propose of being correct that I dare any one to disprove any part or the entirety that my concepts about cosmology forms! But that can only come about when I can get an audience to see how I expose Newton for what Newton was and it is in that where I find no luck. I can't find one academic with influence that is brave enough to stand up and face my attack on Newton and argue me down or prove me wrong in a sound debate. Now I see frowning coming from everywhere because it is madness on my part to think the world is wrong and only I am correct!

I realise that it shows signs of madness on my part and in my thinking to even regard any possibility that I am the only person on Earth that is correct and all others that ever studied physics are wrong but mad as it seems, if that is what I have to say to find an audience to listen and to judge my case, then that is what I say. I don't say this lightly or without understanding the enormity of what I suggest is going on, but be that as it may seem, it is the truth without question that Newton went on for three hundred and fifty years defrauding science with no one testing his claims. Argue me down or prove me wrong but don't discount me before hearing me out and only after considerable consideration while studying my arguments then form an opinion that disputes what I say but when disputing what I say, do it while confronting me in a sound argument when proving me incorrect! This not one academic could achieve and I challenge the lot to do so. But do it after studying all my work and being in a position to account for all the details I propose. Don't just dismiss me because I dismiss Newton because following that road is the way of the coward and the mentally impaired. Read my challenge about the correctness of Newton's proposals when he brought no more than suggestions into science and when I dispute Newton, then take me on by proving Newton correct... do it just once... prove Newton correct just once...prove that his formula is working and that his principles apply on the grounds he principled his ideas.

Detecting Newton's misconduct is possible because I saw a way to break away from the invalid concepts Mainstream physics hold. I went about and tried to prove Newton and when that was not happening I tried to apply Newton's ideas into the greater fields of cosmology. That also wasn't possible. I tried to amalgamate the four cosmic principles applying in cosmology with what Newton said was happening in the cosmos with mass and with gravity and in light of what the cosmos showed was happening Newton just wasn't happening!

Notwithstanding the pose Mainstream physics try to uphold, the entirety of physics still use the idea of magical forces intervening in nature and they still base concepts on unexplained novelties. Think of finding four unexplained forces going around and influencing persons in an unexplainable manner except that the magic of gravity keeps people attracted to the Earth. To say the least, the concepts physics use in terms of Newton would not even be acceptable to children in the modern informed era we live in, I challenge any person to prove Newton, not to accept Newton but to undoubtedly prove Newton correct! Prove how Newton's formula of mass forming the force of gravity can apply as Newton said it does! I recognised the impossible double standards Mainstream physics apply to promote their much shady explaining. In short I tested Newton's principles and found the principles to be wanting.

The inconsistencies Newton introduced brought science double vision and to compensate for these bogus truths supporting their incredible theories, they simplify issues to such a level where what they embark on, is the meaningless acceptance of the unproven and they proclaim to understand what are meaningless inconsistencies and to achieve this they create scenarios which uses the entanglement of deception. Prove the attraction Newton said was enforcing gravity that is pulling by mass and is gathering plants by contracting the diameter between planets.

Show how much the Moon came closer to the Earth since the time of Kepler. Show proven distances taken by radar tracking and indicate just how accurate Newton was. Show how much the Moon came closer to the Earth since the time of the Moon walk in sixty nine. The figures are available but are kept in a grave of silence where no one ever speaks about what science found applies and how much the distance between the Earth and the Moon is shrinking as Newton said is happening or then how much is the is expanding which will contradict the very principles Newton brought about! What they declare as unwavering facts can't even be supported in some form when tested by a silly test as to show that the distance between the Earth and the Moon is shrinking. Even the least degree of verification of correctness is absent when trying to find support of Newton and Newton lacks all evidence of authentication in any investigation of even the simplest terms. It is as if they never read with interest that which they explain when they embark on explaining Newton and they never scrutinise that which they advocate when they teach Newton's principles applying. They give values that are senseless and the very values they use make that which they say meaningless.

In this book I am going to investigate how much truth there is in mass pulling by the force of gravity. To most if not to all of the persons reading this, such a venture of investigating Newton is time wasted and just the thought about me embarking on the investigation of the issue is totally senseless to investigate. It is senseless because the concept it carries became accepted as household practise and life science from where it proceeded to become everyday culture in every person's mind. The worst part is that the group of people normally considered as the wisest bunch there is, never did prudent testing on Newtonian presumptions, while to test the presumptions is most easy to do. I will not believe that a lot that lives up to the veneer of being the best mathematical intellectuals on Earth, never though of testing Newton's very simple formula and in that disregard the formula because of the incorrectness the formula holds.

Do you think of astrophysics as being the department that is run by the wise and the level minded, the honest and pure at heart, the nobility of well-to-do academics and the sober thinking standing in front of the world as the absolute trustworthy? If you are a student, there is no other choice you have but to trust them while they feed you absolute hogwash! If you would so much as dare to doubt any thing they say they will banish you from the institution they rule so absolutely. The banishing process is dome under the blanket of examination. They teach you what to think and to make sure you think what they wish you to think, they tell you to confirm their teachings on a blank piece of paper. You write what they prescribe and you supply the answers they demand in the words (sometimes) of what they demand. Should you in any way say anything different from what they tell you to think, your presence will not be tolerated any further as they abolish you from their institution of academic tutoring.

After reading this book I invite you to...no I dare you to challenge their statements with evidence gained from this book and see them wilfully further their culture of deceit by bringing unfounded arguments just in order to silence you and prevent you from getting behind the truth. If you think those in charge of astrophysics are the pillars of trust, then get wise by reading the following facts and arguments this book presents. What you are about to read is simply mystifyingly simple and yet to this day I have not had the privilege to challenged one academic any where that had the honesty to admit to the fact of Newton being wrong. After you have considered the following you might agree with me that even small Children can reach a higher level of clear-minded logic and find more sensibility than what those scientists promoting astrophysics have because science lives in a make believe fool's paradise.

The manner of regard to life that the Academic Physicist holds and the outlook on life that the followers of Newton physics have (I call them plainly Newtonians and to me they are sheepish because they resemble to the image that to me seems the same as sheep running after their leader without having the ability to think for one second any thought spawned out of personal intellect) is quite the opposite of what I think of them. They keep their forming the establishment of the order the Academic Physicist in high regard and consider their order to be the top thinkers in society.

This religion that they practise of self promotion and sublimely self regarding their status being next to God has them so high that we down on Earth forming the waste of human garbage can be told anything and we will believe what they say just because they with their supreme intellect tell us to think what they wish us to think. This they do because we human waste living way down below their supremacy have not the ability to think and therefore they must think on our behalf. In their view and so far very correctly judged on their part, they, the persons being in the group that forms the Academic Physicists, believe very correctly that can dish up whatever they wish and we, those forming the group in the gutter, those that are mindless in their eyes, we will have to accept what they say without being allowed to form an opinion other than having the opinion they give us to have because in their view we are unable to have a mind other than what they are able to control. This attitude they have is the result of a relationship that worked for so long and thee fact hat it worked that long is what confirmed their opinion that we, the public, are fools to believe anything and everything because of blind stupidity.

But in spite of their aggravating conduct and mischief towards us, it is not because of a lack of insight and inability of controlling a mind that we have our childlike belief and blind trust in their opinions and which there was. It is the faith we shown that they misused for their scandalous cheating. Our faith is what we have shown towards them and is that, which became used as the reason why we accepted what they said blindly. We didn't accept their word on the grounds of us being utterly stupid as they perceive us to be but our trust depended on our good nature and believing in their trustworthiness.

This trust we have is brought on by a culture of trusting the King to do the people well and somewhere in every person's cultural past there was Kings that did us well in leadership. But their underestimating of our abilities is the testimony of their poor understanding and their weak insight ability, which results from their arrogance and stupidity. You are about to see just how stupid they really are in the thinking aspect of science. It will become clear as you page along while reading! They didn't fool us half as much as they fooled themselves and you are about to read all about it. The fact that they could fool us for centuries didn't run on their intelligence being so much superior but served their purpose as it stemmed from the trust we had in them resulting from good intentions on our part. This betraying on their part and misusing the public's good nature to be used in schemes to get the public conned must end and I pray that this book form the first step in resisting the arrogance of the Academic Physicist.

Any one not in their group of the Academic Physicist is part of the lowest order of mindless being and to become part of their order and those that have minds with an ability to think, students have to accept what they say when they say whatever they wish to say without having to prove the correctness of what should back their saying so and as a result of this students may never question what they say. Only when and after proving that a student has totally lost all ability to think for him or her self may a student be promoted into the ranks of their sublime intellectual group. The sifting process they named examinations. You write on paper what they told you and never question their opinion and after passing that examination will you ever enter their sphere of intellectual brotherhood. Does this sound far fetched? Then you better read on and I will remove your blindfold and show you what a world of deception the Academic Physicist force on us into.

Read the following and see how they, the high and the mighty, those that think they can replace God and those who think they can think on our behalf and think what to tell us to think, how much they are clowns and the jokers in society. Read how little are they, the Academic Physicists, able to understand concepts about Creation while they think they are able to replace God in their superior intellect.

If you are a student in the science of physics, then ask your Educated Masters to please explain the following abnormalities you are about to read in this book and insist on a clear explanation about the inconsistencies they promote while tutoring physics as if the physics they present are the most flawless and accurate institution there has ever been. Ask those academics supporting Newton about the following flaws that no one mentions …ever… except me in this book you are about to read and get them to explain the inconsistencies never talked about, which I present in this book and then after confronting those charged with tutoring physics and seeing who should be believed, then get wise instead of brainwashed. Let them mathematically show how one would go about and use Newton's visionary formula $F = G \dfrac{M_1 M_2}{r^2}$ to calculate the force of gravity by replacing the symbols with the actual values in mass that the items referred to have. Put in the Earth's mass in place where it belongs and put in your mass in place where it should be and then divide that with the distance between your soles and the Earth measured in micro millimetres by the square thereof!

In the book named an **_Open Letter on Gravity Part 1 and Part 2,_** I bring the solution to the mystery behind gravity. I tried in vane to introduce the principles I find valid to the academics in charge of astrophysics. Facts that Science present as being the uttermost explicit and unwavering truth, fails to bring any logic answers to so many questions that it should address. It fails to have substance in addressing the most basic and simple questions about gravity and physics. Yet to every question science can't answer my approach does bring many solutions. The presentation and the delivery of my answers that I reach are understandable and simple where it serves both logical science and the truth.

Since my answers do not match Newton and his misconception about gravity and that mass generates gravity, those in charge of science don't even bother to read my work. With their affixation to the corruption they portray I can do little to the giants where they are in the mighty positions they have and just because of that they can go about to sideline and ignore my work and this is notwithstanding the correctness that my work delivers compared to the utter failing that Newton's work shows. When confronted with my evidence and they have to match my work with the hypocrisy and misleading nature of Newtonian cosmology their defence in substantiating their claims is to ignore me. Since I do not applaud mainstream science and the clear fraud they embrace and fraud it is that they embrace, I am silenced.

Why is it that my work is going unrecognised or even in the least goes never debated and never commented on…it is because it will then trash every article anyone has ever written about astrophysics and cosmology. They show little integrity when academics with such supposed high standing or then such as they should have, play a dishonesty game where those in commanding positions will rather protect fraud and save their skins. They would rather protect the corruption they have than seek the truth and find honesty in physics. Those academics in charge would much rather protect their un defendable ethos they maintain as forming the back bone in science and what gives their personal position legality although it is corrupt than admit to the truth they find when they begin reading my work and in agreement they then have to back the truth my work brings.

Doing that (accepting the truth in my work) will trash all work in cosmology delivered thus far and condemn it to the waste paper basket and render all work invalid and void. It will put all the Newtonian's bias and fraud into the place where it belongs. Considering that such acting will lose them money, those academics in controlling positions then will rather rape the truth in order to benefit from continuing to corrupt student's minds further. If they wish to justify their inconstancies they have to attack my work and disprove the accuracy of my work. That they can't do. They then ignore my work because they can't attack my work. In that sense they also place their work beyond my approach, as they can simply ignore me as if I represent the plague while they carry on with little consequence to bother them. I challenge them to prove Newton correct and not just declare Newton being beyond reproach after all has seen the evidence I bring. After reading this all students must challenge them to defend what they can't or get honest.

$$F = G \frac{M_1 M_2}{r^2} \quad F = \frac{r^2}{M_1 M_2}$$ This is the basis that Mainstream science uses as the foundation of all physics anywhere. If this is wrong then everything they have got to work with goes out the window. They put mass and the distance that parts objects in a relevancy, in other words the one is a ratio to the other. The one factor brings a measure to the other factor's value. The one cannot be without the other. The increase in one becomes the reducing of the other and the other way round also applies. When the distance is large, the influence of mass will be small and when the distance is small, the influence of mass will be overwhelming.

Then they state we are in a Big Bang expanding of the entirety. Why then, when considering that if it is mass that produces an inclining force of contraction as Newton says there is going on then…why didn't the expanding stop before it started when the Universe was small. Today using hindsight after the fact of the exploding Universe became apparent by the studies Hubble brought to light did the lot of everything that is not implode as Newton would have us believe whereas, instead it did expand just as Hubble proved. The radius at the time of the first instant back then was no factor, which makes the gravity at the time a totality of unrivalled force. The radius being that insignificant leaves the mass unchallenged in asserting power in relation to the non-existing radius it had.

I dare any physicist to show me where they apply Newton's formula just and exactly as Sir Isaac Newton suggested gravity applies. Show me just once where the mass of the Earth is multiplied with the mss of the object in normal physics. Show me just once how $F = \frac{r^2}{M_1 M_2}$ or $F \alpha \frac{M_1 M_2}{r^2}$ where one M represents the mass of the Earth while the other M represents the mass of the object and in this formula the end result will have a value of 9.81 Nm/s^2 … show just once one example… where the use of the mass of the Earth comes into play. If multiplying the mass of the Earth with the mass of an object and dividing that with the distance parting the two mass factors does not deliver 9.81 Nm/s2, and then any claim by Newton indicating that $F \alpha \frac{M_1 M_2}{r^2}$ is equal to gravity, such claiming constitutes to deliberate fraud…even if Sir Isaac Newton said this. Prove that the mass of the Earth with the mass of an object and dividing that with the distance parting the two mass factors delivers 9.81 Nm/s2 or admit physics is conducting fraud to protect Newton!

To whom it may concern:
My introduction as well as introducing the readers to general cosmology in a very brief and compressed manner but first, I have to give the emphatic warning to all prospective contemplating readers.

Please take note of a conscientious warning about the gravity of the misgiving there is on the part of the most respected Academics in physics about a much concerning matter. I state it emphatically that science accuses me to be not schooled to the point where I am able to have any form of an opinion on any matter concerning Sir Isaac Newton. Notwithstanding that my research proves I did my private studies and through which I skipped the indoctrination and mind control academics place on students goes unrecognised by their standards and so too my ability to have any insight on matters regarding physics. However my skipping their methodical and systematic brainwashing enabled me to see and allowed me to be able to express the incorrectness in Newton's teachings and allowed me to show in clarity what destructive force Sir Isaac Newton used to corrupt the laws of mathematics, corrupting to science along the way and mostly raping to the work of a great man, Johannes Kepler and what Sir Isaac Newton did can only be expressed as being blatant criminal fraud. What his deeds amount to is to corrupt the laws of mathematics, to render the laws of cosmology useless and to rubbish all of science. Should you find this to be unbelievable, then I am glad to announce that this book is more for you than any other person, so go on and read what academics guarding science never wanted published. I challenge any one that disputes any claim I make to prove me wrong by proving me wrong and not merely suggesting claims in that direction.

Tell me, can you find any credence in the "Conversions for "Unknown""

$4\pi^2 a^3 = P^2 G(M + m)$

In this comes the fraudulent part because there is no evidence of mass playing a part or forming an actual presence in the solar system.

If the cosmos supported Newton's claims of $P = \left(\dfrac{4\pi^2 a^3}{G(M + m)}\right)^{0.5}$ then the planet arrangement would have been much more likely as I show above, but the picture indicates the mass as well as the planet formation.

You must judge; it is either the cosmos that is incompetently wrong or it is Newton that is incompetently wrong because what the cosmos has in place Newton knows nothing about and what Newton claims the Universe uses, the cosmos knows nothing about. Who would you say knows more about the cosmos' method of workings, Newton or the cosmos? If Newton is correct then the planet layout must be as I show with Jupiter very close to the sun. It seem the cosmos is just as unaware of Newton's ideas as Newton is of what is happening in the cosmos. Who would be correct about cosmic principles applying, the cosmos or Newton?

$P = \left(\dfrac{4\pi^2 a^3}{G(M + m)}\right)^{0.5}$ What hogwash does the factor $\dfrac{}{G(M+m)}$ indicate?

The same can be said in the formula $M = \left(\dfrac{4\pi^2 a^3}{GP^2}\right) - m$ when $P = \left(\dfrac{4\pi^2 a^3}{G(M + m)}\right)^{0.5}$ that the factor $\dfrac{P^2}{}$ is senseless and $\left(\dfrac{P}{2\pi}\right)^2 = \dfrac{a^3}{G(M + m)}$, has no foundation other than fraud.

$M = \left(\dfrac{4\pi^2 a^3}{GP^2}\right) - m$ is complete fraud. The Cosmos does not support the Newtonian formula even in one place where it could apply.

Position as a function of time

$P = \left(\dfrac{4\pi^2 a^3}{G(M+m)}\right)^{0.5}$ This is what Newton said is in place and with no evidence ever founding this ridiculous proposition, all Newtonians that ever come after Newton. This is what Newton and his Newtonian followers tell the solar system it has in place and tell the cosmos it uses to operate. I have indicated that mass has no place or use in the solar system according to what the solar system puts in place.

These are the closest because these are the massive giant gas plants and having the most mass must put them the closest to the Sun.

Get your professor to prove Newton correct in the face of $P = \left(\dfrac{4\pi^2 a^3}{G(M+m)}\right)^{0.5}$ and if he can't let him admit he has been conducting in a fraudulent practise all the time he was teaching.

IS THERE DEATH AFTER LIFE?

Science believes in science with no reservation. Therefore scientists believe in science with a belief more that Theologians could believe in God. To Physicists everything there is must be science. They formulate a mathematical equation and when doing so the explain everything in man's field of knowledge. Those in science propagate that we will live up to a thousand years in a very short time in the future. However ask them to explain aging and they have no idea what they are talking about. Science puts life down to some acids and a jolt of electricity and with that life could be programmed. According to science life is mostly electricity stored in the brain and by electricity the body moves life around through the body. What a lot of hogwash and this attitude of simplification is so Newtonian junk as Darwin's simplified ideas are about the origins of life. To any Newtonian any suggestion in a simplified form regardless of proofs everything science requires to become accepted as fact. The overwhelming majority of physicists with Doctoral degrees in physics will not have the ability to read this article and too understand the arguments. That is because there is a shortfall in their argumentative ability. This inability runs very deep in physics.

When there are aspects in nature that physics can't address then we have to look for the shortfall in physics. Physics holds the opinion that God cannot be proven or be substantiated. The fact that physics can't accommodate a certain fact or feature does not exclude the fact, but it underlines the failure in physics. Prove a thought using physics and see physics fail. In example when we look at the fact that I can think and my thinking is a fact beyond proof, yet physics can't prove how I think or why I think, except put it down to a flow of electricity in my brain and body. Other than that they have no capacity to know what life is or to understand the concept of life. If they shock a frog leg with electricity and the muscles show spasm they then conclude life is electricity. It never dawns on them that life is the ability to generate electricity whereby muscles are controlled and life is a lot more than just the flow of electricity. This shows the absolute fallibility of physics and not the absence of my thoughts. The Newtonian mindset is to keep physics above reproach and beyond suspicion while it is desperately poor in all senses.

If physics tries to put my thought down to some brain activity somewhere in my mind it shows how incompetent the reasoning is behind the argument. The activity is electrically induced and could be harvested by jolting on nerves from some exterior source and that does not prove that my thinking comes from the brain being jolted by electricity, it proves I am some thought with the ability to jolt the brain into action by supplying the electricity. As the medium could arouse action by stimulating the brain with electricity, so the true "I" has the ability to stimulate the mind with a jolt of electricity to provide electricity. If it was the electricity that did the job, we then could jolt a cadaver's mind and get the cadaver filled with life. The brain is not what stores life but it is life that keeps the brain with life. If life was in the brain we could generate the brain back into "life" by shocking the brain with electricity hours after death.

Once life is lost, no electricity can restore the factor we call life. Life we know is in thought and thought is the presence that physics never can accommodate and yet the entirety thought to be life in whatever form is held by thought and with thought the body is motorised by life. Yet, when physics can't accommodate thought we do not discard thought as an absence that is unproven. My thoughts are with me until I die. If physics fail to accommodate God then it is not God not being a reality but it is physics being an utter failure. I can and I do prove God as the absolute factor in the cosmos, but before one can get there a lot of Newtonian garbage has to be discarded and a lot of misleading Newtonian disbelief has to be dismissed. If you look at the night sky you will see many specks of light. In that understanding but more advanced to understand than any person can realise we find the proof of God.

It is not in the light but in understanding the principle forming light and the understanding behind the concept of light that we can mathematically prove God by using mathematics. I have done just that but it takes a book of 700 plus pages just to explain the reality forming the concept. Atheism is not proof of intelligence but it is proof of the lack thereof. My dog is the biggest atheist walking this earth but he is that because he is stupid and has a lack of mental capacity. One must star to understand light and understand the light is the Universe. The

Universe is not material and darkness but it forms by light. One doesn't see a galactica, one se light that formed a galactica a very long time ago. That what I see is the Universe and the Universe forms by light left behind by time as space. I am not going into that because I wish to try and keep it simple.

Have you as you sit reading this part at this minute sat back and gave a thought about the light enabling you to read? Such a thought brings to mind the most simplistic answer one can imagine. The light hits the page bounces from the page and contacts the lens of my eye where the lens conveys the photons becoming electricity to a part of the brain that translate the electricity to an understandable message and that makes one read. It is as simple as that! Ever gave a deeper thought about light streaming across the night sky, coming from the visible limits we think of as ends of the Universe we do not even realise it is there? How does the photons manage to convey one complete picture coming from as far apart and as wide an area as it does? With a few photons connecting the eye or lens no one ever noticed the wonder of light. The photons reflect a view that seems as if coming from all the billions upon billions of stars. But most is coming from darkness covering an area no man can measure. Yet how many photons can actually connect to the lens of the camera or to the eye considering the size the eye allows light to pass through? We see by using a few photons.

We see with a few photons going past our lens a Universe representing immeasurably many photons. Still a few photons coming from a single direction directly ahead eventually tell the entire storey of a Universe larger than we could ever understand. What we see is bigger than any person can comprehend and what we apply to see that much is smaller than any human mind can comprehend to understand. It is very simple to take the process of seeing by means of photon conducting very lightly and I have never heard one of the Brainy Bunch really in sincerity uncover the process to its utter and full potential. Moreover let those Mathematical-Masters put this notion into an equation and conduct the understanding they have by proving the concept I just explained. It is impossible that light from such an array of assorted sources can simply come together at the eye lens and show a picture of objects spanning across a Universe as wide as our mind can receive where the objects they reflect is beyond human measurement and the quantity we can apply to receive a vision is inconceivable many. With that small space within our eye how can we see the space that is large as the Universe that we see? I am never going to try and Simplify the idea behind understanding God in terms of physics, but very deep inside this understanding we can begin to fathom the presence of God in terms of physics, not religion according to a Bible, but in terms of Physics. But understanding that is way ahead and light-years advanced from what I explain in this book.

Light is much more than the medium science takes it to be. Light connects the Universe in a way we cannot contemplate. Light being far apart originating from regions not in the same time or Universal space connects in a way that present us with a picture holding the Universe in an understandable content. From the point we stand and we watch the Universe the significance of what we see surpasses the sense of understanding of what we are experiencing and more so surpasses what we are able to understand. How can the few photons that our lenses catch coming from such an area as the night sky cover transmit the complete picture of what we see. Take a few seconds and study the picture of the night sky then rethink the picture applying the full content in the picture to what the size of you eyes is. Think how big the picture is that your eyes take in and translate that area to the size of your eyeball in an effort to determine a ratio.

One will be forgiven if one thinks of the ratio as eternal to nothing. Yet a few pages back I showed that according to mathematics there couldn't be anything as nothing. Consider the path the light followed from the source connecting to light from all other sources where all particles of the other light may come from and bringing a full picture to the lens one use to look through.

In your mind connect a line from every atom producing light and connect the lines to your eyeball and see how you can manage to fit all the lines, as small as the lines may be. Understanding this can only come when we understand singularity and I still have to find a person educated or otherwise that would be able to understand this idea.

If it is lenses that enable us to see what we can't see in outer space it also means we cannot see the light, which is outer space because we haven't got the lens to match the curb of outer space. Newtonians think of outer space as geodesic zero, with nothing in outer space but space. Geodesic zero means the light travels in a straight line from where it originates unhindered all across space to where the light connects the eye. Such an idea by itself is outrages because the stream of photons reduce in space to such a minute quantity that taken the area the photons travel and the space in vastness it covers, the chances of one photon coming across many hundreds of light years through billions upon trillions of cubic kilometres of space and selecting my eye to convey the electricity is less than infinite. Yet such conveying takes place every second of every minute. The position of the location of the second singularity, which is the precise duplication of the first singularity but in a diminished capacity, is obvious to miss when one is not applying a detective mentality, as one should in scrutinizing the cosmos. Culture will have us believe that when one sees a colour shining from an object the colour is associated with the object. Logic tells a different storey. A yellow dot is all the colours in the spectrum but yellow because it is disassociating with the yellow. That goes for red blue and all other colours we may visualise. I think the norm accepts this as scientific fact with very little argument or substantiating proof about that required.

If light came as individual streams of photon flurries, then our visage would translate that as such shown in the fragmented as telescopes enlarge images. If the light only held what we think of as light, we would be unable to see the dark parts because only the light parts would contain light. The total picture we could see would be a picture unconnected bringing across some photons in the manner where every object stands apart not being related in any way and that will be what we see, if it is anything that we see.

That we know is not the case but that means geodesic zero is as much rubbish as anything Newtonians regard with simplicity and with careless thought. Geodesic zero means nothing and how can I see nothing as darkness because "nothing" is not darkness, nothing is "nothing" and the darkness I see is darkness showing the darkness as something. The darkness we see is as much "light" as the "light" we see because we see Light" and in that we see "darkness" as another form of light.

What then about colours that are technically not colours as is the case with black and white? White is simple. By spinning all the colours in the spectrum the colour white shines through. Black is quite another matter. A friend of mine whom is one of the best painters I have ever come across told me that one couldn't paint black but have to make black a dark blue to show shade on the canvass. That apparently is his success in achieving the realism. He also went on to explain how many variations of dark blue form the shadows in one simple tree. This remark set my mind in motion. One cannot see black because black has no colour to show, but black is the colour most prevalent in the universe. One can see only by colour and since black is not a colour we should not see black, but we do.

The fact that we see light means that the dark next to the light cannot be "nothing", If the darkness was the representation of "nothing", then that should be exactly what we must see, nothing but the stars. Taken from the top picture some stars and leaving the rest to nothing is what we see in the picture below. A blind person sees nothing but when we look at space, we see something that we think nothing of as we see as space. One cannot have the ability of sight and see nothing. It is light that we see and it is light that we use, which enable us to see.

That proves the darkness that we see in outer space is light that we see without recognizing it as such. If the darkness was the representation of "nothing", then that should be exactly what we must see, nothing but the stars.

Taken from the top picture some stars and leaving the rest to nothing is what we see in the picture below. A blind person sees nothing but when we look at space, we see something that we think nothing of as we see as space. One cannot have the ability of sight and see nothing. It is light that we see and it is light that we use, which enable us to see. That proves the darkness that we see in outer space is light that we see without recognizing it as such.

What puts us humans in a category one higher than animals (or so we like to think) is our ability to think about that what we can see. The less develop an animal is the more it has the attitude of eat or be eaten. The higher developed animals are the more the animal find reason to argue. One may teach a crocodile not to eat you if you start feeding the animal. That is a mindless reptile and yet it can think above eat or be eaten. What we see is not merely the truth and it requires reasoning to see the truth and substantiate between culture motivated observations and thought through decisions.

To all the **Super-Educated-Mathematically-Superior-Intellectuals** physicists believing in mathematics and those trying to replace God with mathematics, prove the ability to live by calculating thought as the sole factor of life where it is life that controls the body and not the other way around. Should any of them insist that the mind is responsible for life, then revive a cadaver by filling it with whatever acids you claim produces life or shock the corps until it roasts or force movement onto the body.

If it can't revive, then go and calculate by mathematics the precise ingredient it is that has left the body and therefore has filled the body with death. Mathematics is as unaccommodating to reality as Newton is to cosmology. All those Super-Superior Newtonian mathematicians, use you Newtonian inclined mathematics to explain my previous argument about how all the light that fits and fills the entire Universe can bring one picture of the entirety to fit into my eye.

The fact that **your** Newtonian physics will not allow you calculate this does not remove my ability to see the entire Universe in as large as it is through one tiny hole at the back of my eye. When you understand this entire concept you will have the ability to understand God's presence in the Universe and until then your mathematics removes your ability to understand physics and Newton promotes you blind stupidity about real cosmic physics.

Moreover, I prove all of this ability mathematically but only after removing the falseness of the factor of mass and from the myth presented as Newtonian corrupt science.

This article is as much about proving what energy is as it is about knowing the difference to the state in which alive person is and in which a dead person is. Newton considered all forms of energy to be the same, and oh boy, was he mistaken. It is not surprising he   formulated gravity the way he did. There is a worldwide fashion amongst the very well educated that in order to be regarded by those with the know how as a supremely informed person, one must at least be an atheist. The key to science is apparently to be completely atheistic. Atheists do not believe in the life after death, a Creator or a Force that does not exist outside the technological criteria of mathematical science.

Everything that does exist only exists because it exists in the perceptibility. Any force that might lie outside this norm is quite unthinkable and that thought could never present itself as to be present in the material universe. The ironic of this fact is that those well-educated scientists have only one source of information and that is light waves. Still they permit themselves to be atheists in their blind state of ignorance.

I do not condemn them, because they apparently know more than I will ever know. However, because I am not that knowledgeable, I must feel my way through the tunnel of ignorant darkness like a blind person. However, in doing that, I stumbled across a heap of questions that has no answer, even by those who carry the flame of knowledge. That forced me to form my own theories, think and come up with sensible conclusions, which answers all those questions their light of knowledge could not answer.

I declare to be of average intelligence and like millions of others on earth, all these millions are believers, like me, and are confronted by the same questions these super intellectuals are seemingly incapable or unwilling to answer. Then I realized the super intellectuals only have one source that lead them and that is measured light.

Let us look at the definition of energy. Energy is, as I understand it, indestructible, which means it cannot be destroyed. Energy can only be transferred from one form to another form. Let us look at the example, which is used to teach scholars at school. We take a rock and move it from a ditch up a hill. On top of the hill, we have a lot of potential energy that was transformed from static energy by means of kinetic energy. This is the simplest example we teach children in school.

I too had to teach the children this nonsense, in the period when I too was a teacher. In the transformation, other losses occurred, like heat, sweat vapour and friction losses.

The science apparently does not take into account energy losses brought about by anger, fighting and frustration brought about by incumbency. These are also energy losses. After all the sweat and wrestling, the rock is on the top and we have a situation with potential energy from which we can derive kinetic energy when the rock is rolled down hill. I do not agree with any of the above mentioned, and will later state my point of view. I will however declare at this point that Newton's statement of energy and work being the same thing is utter nonsense. If I feel drowsy it be because I lack energy and if my car stops running it is because of a lack of energy and if my dog bites someone it is because of too much energy

Newtonian physics can't prove a God and therefore there is no God. Newtonian physics can't prove any form of feeling or emotions so we have to accept there is no feeling or emotions just because Newtonian physics are incapable of proving we have feeling or emotions. The prove or no proof Newtonian science can comply with is incredibly limited although Newtonian science wishes to cast the view that what is not Newtonian is not reality while the fact is that Newtonian science is not reality.

When medical doctors have no idea what is wrong with a patient it is because he smokes or because someone next to him in a bus twenty years ago smoked or if he never saw someone smoke even on Television twenty years ago and you saw it on the TV screen it gave you your heart condition and if that never happened then the person has cholesterol because cholesterol takes all the other blame smoking can't take. That is the way all science are run in order to keep the charade that science in any form is infallible. If they don't know what is wrong it is either smoking or it is cholesterol but it is never that they have no clue why you are medically the way you are because they have to pretend to know everything. This is science. If science has no clue what they talk about the use the word energy. Energy can be anything science wants it to be because…science says so.

Life starts of being in a sperm that has to couple with an egg. The sperm only carries life but does not even represent life. If life leaves the sperm you can do to the sperm whatever you wish and it would not represent life. Therefore the sperm is a vehicle for life and life forms the sperm sell. If life abandons the sperm sell the sell goes back to atoms. If it were the other way around, the sperm cell would remain intact and start hunting for a new life form to hold. The same argument applies to the egg. Thee egg carries life and life supply the other half of the life that will become human. If the egg does not hold life any longer, then the egg will disintegrate into billions of atoms once more.

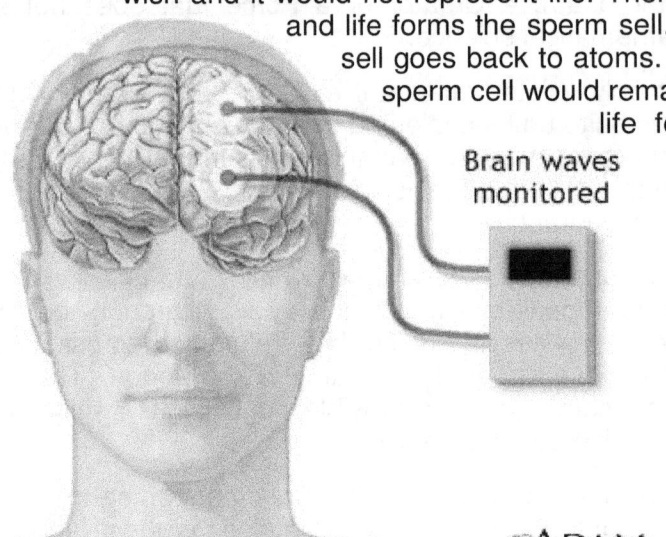

Brain waves monitored

This is extremely important to realise that from the first second of life forming life collects tissue that will become a human body. It is not like your halfwit Newtonian professor believes that the human body represents life. From the first moment life forms the body and it is not the body that forms life. Therefore you with your life forms your human body and it is not your human body that takes the responsibility for life.

It is by the thought process that life collects material to form the human body and the human body does not collect life as it goes along. Every one in modern science think it is the brain that controls the human body but are they so completely wrong. You use your thought process to control your body and in this thought process you form your body to be as strong as you wish it to be. Can I make you strong; no I cant because you are already as strong as a giant. What I can do is help you realise you potential strength by helping you learn to control your muscles. Before you complete any action of movement a thought first have to apply. It is the thought that command the muscle and not the brain that commands the Muscle. The thought takes charge of a cell in the mind and then takes information stored in the cell of the brain, which a thought directs to a channel that by electricity which the thought is also responsible, inflicts current in the muscle to pull the muscle. The thought generates the electricity that collects the information and the thought sends the electricity carrying the information to the muscle that has to do the job. It is the thought and not the brain and there science falls flat in their hogwash they use as information.

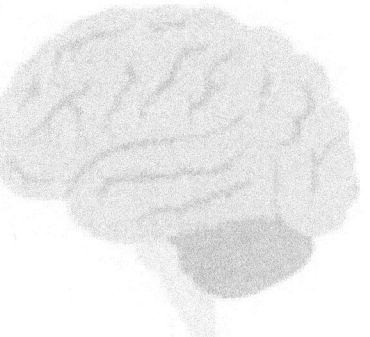

Life which is what you are, not a decomposable body, started accumulating material by thought when you were sperm and egg, and after the Unification you started accumulating useful building material. It is done by mind controlling the body. Don't allow the atheistic senseless stupidity tell you different. If your body was what is in charge which is you, then when you are dead someone with life can pump some oxygen into you and shock you with electricity until you bounce around like a ping-pong ball and you will

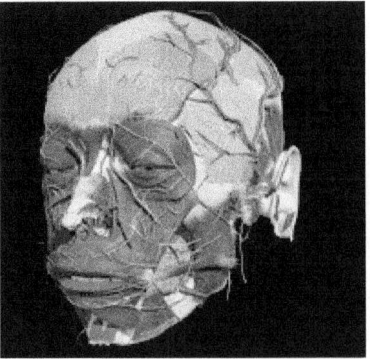

begin life again. That is total rubbish. If life leaves the body there is no structural formation left to control and maintain the structural I integrity of the body.

Even in the very beginning life formed the sperm and the sperm did not represent life. You can't have a tube filled with sperm and when you find the lot are dead you revive the sperm with an electric jolt. You can't have a jar filled with D.N.A and by regrouping the composition you build the body of the person once more. You build your body through thinking with your mind. You construct your body cell be cell by using your mind to do so. How can I prove this? The instant your life vacates the body; the body's ability to restructure the structure leaves that very second. The moment life vacates the body, the body degenerates by fragmenting the structure until the entire construction disassembles into forming atoms again. It is life that keeps the body into form and without life the body de-fragments into atoms once more. Your body doe not hold life but your mind by thought controls your body.

==Your medical doctor will tell you your body represents your life and when you die your body dies.==
He will be of the opinion that when your human body dies you have died. The problem with this attitude is that while your human body is still intact, one should be able to resurrect the body by supplying heat and electricity. That is not possible. That is the way one goes about killing people. By electrocuting people on a chair the state removes life from the body ands so giving the body electricity does not bring life to the body. Therefore the body does not use electricity to instate life but by duplicating the transmitting of electricity the body becomes confused and the body relinquishes life. It allows life to depart.

Your physics teacher / professor will tell you that you are what you are because of the body you have. Hogwash I say that concept is and I prove it is rubbish. You build your body with your mind but you control your mind through thought and without though you will not even move a muscle. It is by thought that you tell a muscle to move and it is by thought you tell the muscle to get active and by the same thought you form the muscle which puts the ability in the muscle to form the strength. I prove that this is how gravity forms and it does not form by the pulling force of mass. Newtonians such as your physics teacher or professor wouldn't even read my books in which I prove they (those teaching physics) are all brainwashing students to believe that physics is what Newton said it is. I prove they are submitting all students to mind control in order to force you to believe physics is what they teach it is. If you want more facts to see if I am correct about your teachers brainwashing you and submitting you to cruel mind control you are welcome to go to **www.sirnewtonsfraud.com or another slightly more complicated version I call** www.singularityrelevancy.com **or** another slightly less complicated version I call www.questionablescience.net , where the websites will show you how much those you trust deceive you with science they can never prove.

Those that think they are experts in physics has no idea about physics and I challenge all of them to prove Newton is correct, not to surmise that Newton is correct or to force students to admit and confess that Newtonian physics is correct but to prove it is correct. Prove the formula $F = G \frac{M_1 M_2}{r^2}$ does form gravity by forming a force by the value of mass or prove that the formula $P = \left(\frac{4\pi^2 a^3}{G(M+m)}\right)^{0.5}$ does put planets in positions allocated according to mass.

It is life that allows the semen to swim and it is the life within that allows the egg to be receptive of the semen. When either the semen or the egg holds no life and the egg or the semen is still intact. There is no life forming possibility. The semen does not swim it is life that allows the semen to swim. This proves that it is life that allows movement from the beginning of where we think life starts.

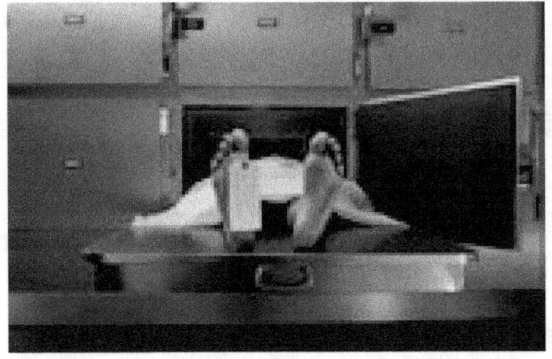
This is a picture of a cadaver. It is not something to be scared of because it is a body NOT containing life, as anyone of us will be someday. So it is the same as you being scared of you as you are going to be somewhere in the future and that is pretty silly. On the condition that you were born the only thing you will be someday is dead. If you are alive then you will face death. What we have to answer is what is the difference between this cadaver in the mortuary and me. One is that the cadaver has no life and I show vital signs filling me with life.

The biggest factor is movement and that movement is linked to thought. Considering the implication of this is vital if you wish to enhance your physical strength and build your body. There are persons in hospital in a coma for years and they apparently show no thought because their muscles don't move and therefore they wither away. The thought gives control over the body and the thought form the muscle and the thought form the size of the muscle.

This cadaver or dead person can't get up and walk as I can. Why can't this dead body get up and walk, it is because the dead has no thoughts. If you think the Newtonian idea is correct that life is part of the body then rethink. I dare you to conduct some tests. If life is electricity as they say it is, then why can you shock that cadaver until it hums like an electric transformer and life will not return? If life is as they say it is electric convulsions then try and shock the brain with electricity and you will find no response. The fact that you can manipulate muscle spasm with electric convulsion shows that life controls the brain by charging electricity and that process is done by thought in life.

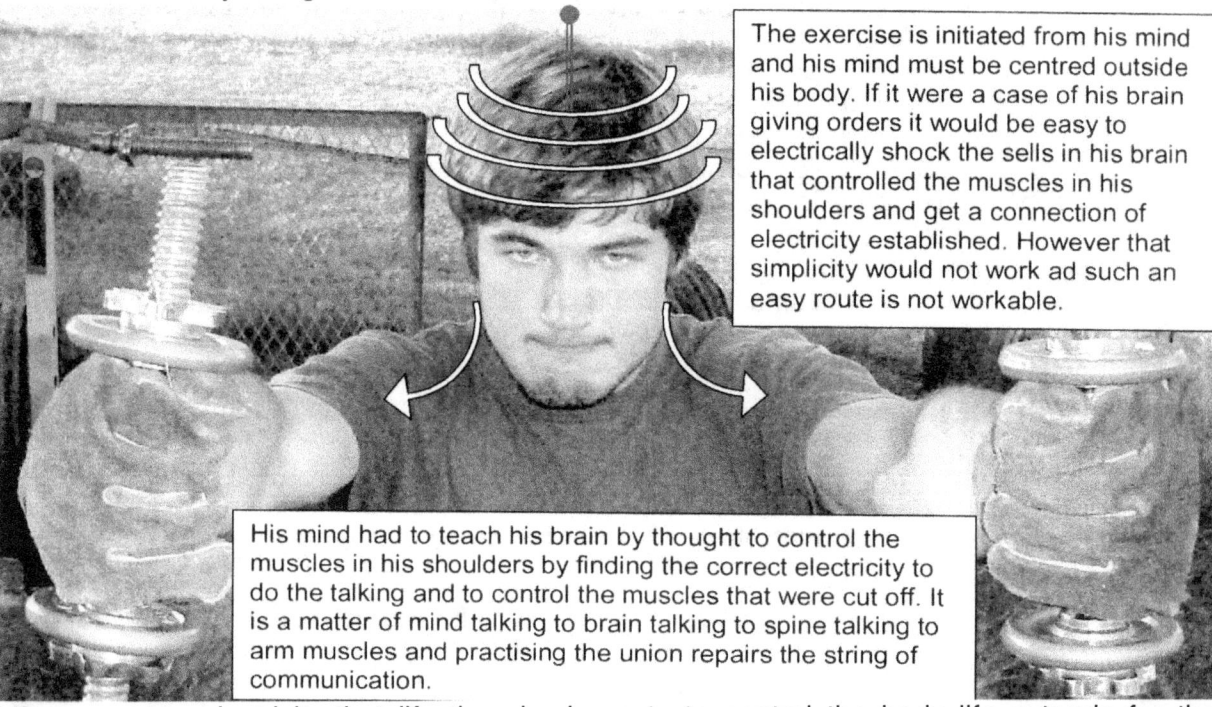

The exercise is initiated from his mind and his mind must be centred outside his body. If it were a case of his brain giving orders it would be easy to electrically shock the sells in his brain that controlled the muscles in his shoulders and get a connection of electricity established. However that simplicity would not work ad such an easy route is not workable.

His mind had to teach his brain by thought to control the muscles in his shoulders by finding the correct electricity to do the talking and to control the muscles that were cut off. It is a matter of mind talking to brain talking to spine talking to arm muscles and practising the union repairs the string of communication.

Life generates electricity that life then implements to control the body life extends for the purpose of serving life.

Life is in charge of the body and of thought and not the body being in charge of life.

By electrocuting a body with life you merely short circuit life's actions with a stronger jolt of electricity but the electricity is just a modem through which life controls muscles and growth in the body. Then you burn the electricity conducting connection that life has as life controls the functions of the body and do that long enough and life may not find a manner to form conduction of electricity whereby the organ control will become suspended.

Your mind charges electricity and that electricity are created by thought and thought is life. By creating thought you form your body and by forming your body through thought you establish you level of strength. That is why when you are in shock you are able to perform in a manner not even you ever thought you are capable of. Around your head there are electricity flowing which science named "brainwaves". These brainwaves are just a form of electricity and that current is the same as what flows around every electrical motor or any planet charging gravity or any star forming a gravitational field.

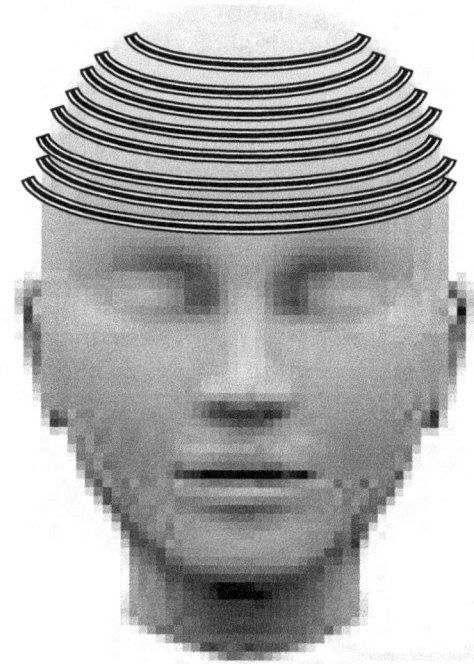

A human is not the body you have but the mind that forms the body

Now we take the scenario of a person's life. When that person is born, and after that momentous occasion of the birth episode, he or she continuous to live on this planet for the best part of the next sixty or seventy years. In this, period a great deal of energy is used to walk, run, laugh, cry, think, produce and reproduce. By doing that, he would from time to time state that he feels tired or without energy. What energy is the man referring to. I have once heard a scientist that made such a fool of him. That scientist declared that if God was energy, God could be coal, because coal is energy as well. Now I would love to invite him to a meal and see him devour a plate of coal. If coal is energy, he can make a meal of it, and then live very cheaply. What he does not seem to grasp, is that there are many forms of energy, which differ totally as we distract the heat and in doing so one can tap the energy. However, coal cannot walk, run, jump and laugh. I cannot even begin to imagine one brick crying and moaning because his friend was thrown into a fire. Coal cannot have sexual intercourse producing an offspring and then caring for it afterwards. Life on the other hand does have that energy quality. This means that there are different values and forms of energy, of which life is one. If life is no different to other forms of energy, God could be another total different concept of energy. This is the problem that I have with these "SUPER- EDUCATED- MASTERS- OF- FACT" geniuses. They can make the most bizarre statements and could be away with it unchallenged.

When circumstances starve the body of food, life occupying the body would begin to devour the body in order to sustain life's ability to occupy the body. To life the body is only a vehicle to serve its purpose. When the body starving, the body does not suspend life until conditions are favourable to have the body reinstall life. It is so very typical of the "SUPER- EDUCATED- MASTERS- OF- FACT" to uncomplicated issues to serve their insight. They make something such as life so simple as to pretend they are completely in control of the knowledge that subject has to offer. The body does not turn life off, life eats up the body until the body is so feeble it can't host life any longer after which when life then rejects the body. The body does not maintain structure but as soon as life evacuates the body, the body breaks down the structure it held when it hosted life. Life maintains the body and will even devour the body and consume the fibre until the body becomes useless to life and until it cant serve life any longer. After life abandons the body, the body returns to a state of atoms with no resemblance of what it were when life formed the body. It is life that constructs the body, maintains the body, and controls the body and consume either the body or as food that is some other life form that had a body.

The world contains a wide spectrum of different occupations that people earn their livelihood from. Seen from my personal occupation, there are two types. Those that farm and produce wheat, corn, barley, nuts, sugar cane, vegetables and many other produce. These are potential energy producers. They produce food, for the other group of the human population that uses this energy product to maintain their strength to apply it to other methods of occupation. Cattle and sheep farmers produce meat that is used by others to convert into energy for their personal use

All people have one thing in common. They devour one form of energy, which is known as food. That is needed to maintain a life cycle, and the consuming of food must be done on a regular basis, to enable a human to live and reproduce for a lifespan of seventy of eighty years. The only precondition is that life would sponge on other carbon-based forms of life, whether it is plants or animals.

This person maintains his way and means of life, thus transferring energy from a form of food to a form of work. Then one day he collapses and becomes still. That person becomes unable to move. We call this state that the person is in, being dead. Even if I take a shovel of food and force it down his throat, he still would lack the ability to transform that energy to movement. But why would this then not bring back life?

Simply because he does not breathe any more. And why isn't he able to breathe? The reason for the person's inability to function, as a human should is because the cadaver is dead. When a person is considered dead, he lacks energy to such an extent that he cannot bring his own body to the grave. Others like me, and I have to carry him to his grave. We, that are alive, and maintain the process of translating food into life, have to carry the dead (he who is without life) to his grave.

The only difference between him and me is the energy form known as life. However, life is not the same form of energy as food, oxygen, heat and electricity. Even if I force all the food down his throat, and pump his lungs with air, while I heat his body with a blowtorch and shock him with electricity, he would still find himself unable to walk himself to his grave. That means the one form of energy is not the same as the other form of energy.

It is widely accepted that there seems to be a generator in the brain that generates electrons which enables the body to function. We know the flow of electrons is due to the process called electricity. On the other hand, do we? In a later chapter, I shall point the difference out between this flow of electricity. However, for the mean time I would stick to this accepted fact that electricity is conducted by the flow of electrons. Now, you can shock the cadaver with electricity until it hops about like a ping-pong ball, if life has gone absent, conducting a flow of electricity would not reinstate life.

You could put the cadaver on life support, with a heart machine a lung machine and all kinds of other machines. This method has nothing to do with life being precious, but fare more with the money paid by his medical aid, being precious. Once the cadaver's line of financial support dries up, his life instantaneously becomes worthless.
Then the cadaver finds the problem that it seems unable to live which means it is dead. Death means the brain is unable to send electronic signals by means of amino acids to the muscles, which would enable those muscles to continue with its normal function. The cadaver finds itself without the energy called life.

At this stage, I think that I pointed out to the difference between a body filled with energy called life, and a body that lacks energy and is called death. However, the energy that I pointed out called life, is miles apart from the energy that consists of food, air, the burning of it and the destruction of it. There is a broad difference between the food process and the actual form of energy called life.

Now I would like to ask those Super Intelligent Atheists and consumers of food and air to explain where the energy form that is called life has gone. Energy cannot be destroyed, but can merely be transformed from one form to another. This is scientific gospel. Life as I pointed out, has a different value to heat. Life cannot be destroyed, that means it has to be transferred from one form to another form, and life itself is not heat, electricity, or food, because applying all those other forms of energy cannot raise the dead.

The fact that energy must be transformed and cannot be destroyed is proved by science to be unquestionable. The life energy started assembling a body albeit sperm or an egg before conception or procreation took place. If the sperm was dead the sperm would not swing and all those that swam in vane died. They did not hang around as lifeless sperm to be vitalised with life as soon as the next opportunity arrived. The very second life left the sperm or the egg without fore filling the process of fertilisation extending the ability of life to assemble more material in order to form a body filled with life, life left the sperm or egg and in that the body holding the sperm or egg destructed. It is life that captures material to form a body notwithstanding how small and in that no one can remove God from physics. The idea that life sprang from somewhere as soon as a sperm was there is as mad as having mass being able to form gravity. With my physics I can prove God being responsible for the flow of time within the Universe and if Newtonian science are not able to accomplish that, it is Newtonian science that is dismally inadequate, but then again that is what Newtonian science is in almost every sense.

The only answer I can conclude is that science is ignoring their own findings to prove their own religion fashions. With life being an undisputable form of energy and energy cannot be destroyed, it seems very unscientific to propagate atheism as a fact.

From these facts, one has to conclude that there does exist another form of life after death.

5. INTELLIGENCE VERSUS EXTELLIGENCE

In this book, I shall introduce you, as the reader, to a completely new line of thought about the science of cosmology. Some of these scientific facts date back to the time when man became aware of a lifestyle that just started to include a civilized order. According to some discoveries by archaeologists, it seems that mankind had its survival mostly due to the way it accomplished knowledge about primitive science, this enabled man to survive in a total hostile environment.

Man's first awareness about forces that he could not control, was explained as forces unleashed by pagan gods. In that is seated mankind's belief and mankind's desire to be in total control of these godly forces. This desire therefore became one of the biggest incentives that drove mankind to a civil obedience and law-abiding standard of living.

We can even today go as far as to except that the role that intelligence played in the development of our specie was far bigger than the role was of the more brutally and physical force. In the animal world, the strongest in specie would ultimately be the leader of the pack. With his brutal power and brute force no one in the pack would dare to challenge the leaderships hierarchy and in that the leader himself. If one challenger should dare to do so, the challenger would pay with his life. It is a well known fact that male baboons not only kill the previous leader, but he will wipe out all the siblings, no matter how much the female baboons might protest against it.

This is even more so in species that has much closer links with mankind. The chimpanzee male just simply murders all possible male challengers until the day he himself is also murdered by his successor. The orang-utan male is another example of a male that would not even tolerate any male in a smelling distance. This confrontation will definitely lead to the death of the weaker one of the two.

There may be a distinct possibility that fear for the unknown was the only reason mankind's development lead us to a higher norm in development than our close relatives. In case of other species the generational development benefits the physical strongest and do not favour the more intelligent of the species. These animals are still much stronger than man is, although man tamed all animals, at one point or another. Therefore, all animals submit to man.

In this, one must define the difference between intelligence and the idea, which I refer to as extelligence

In the understanding of the meaning of intelligence brings to mind how the animal socializes with its own species to guarantee the social survival of the species. That means that all animals have intelligence. Dogs has been with man as long as we can trace back human development, so in doing that he forced the dog into acknowledging mans intelligence. However, the dog still communicates with his own species in the way its intelligence dictates. He sees man as the leader of the pack, rather than a completely different species. When it communicates with man, it will wag its tail or show submission by lying on its back. The dogs intelligence never allowed him to try and communicate with man by using mankind's standards, although his intelligence placed him in a certain advantage point to share to some extend mankind's intelligence. However, the only reason it did so, was to further his own needs in surviving in the pack with man then becoming leader of the pack.

As a farmer, I often watch the manner in which cows interact. At one predestined time during the morning, the mothers leave in a group to have a drink of water. I admit there is nothing strange about that. The strange part is in the procedure, when taking into account that we regard these animals to be thoughtless beasts. One dry cow gathers the entire suckling calves, takes them to a safe, and secured area, where the calves would play and enjoy one another's company. I refer to this as the "kindertiun" which is the kindergarten. After the water drinking, the mother's would gather in a shady spot, and ruminate for about two hours or so. Then they would get up, and stroll in the direction of the kindergarten. Only when they come to a certain point will the calves leave the seclusion, and run to meat their mothers to feast in the generous supply of milk The biggest amazing part, is that the kindergarten hostess is never the same cow. Everyday another dry cow takes on the responsibility of playing stepmother to the calves. Not once is there an incident of one of the calves being disobedient or not under standing what is expected. They always seem to know which cow to follow, and are never fearful of leaving their mothers. They never are obstinate and wonder off in search of their mothers or run to their mothers before someone gives the signal, whatever the signal may be. If cattle are that mindless who decides who's turn it will be to play stepmother, who and how are the calves informed about who to follow, and why are they acting that responsible and disciplined. After all, they are only young mindless beasts. I concluded that in our self-righteousness we under estimate our fellow living species. However, in all fairness to my own species I have to admit, mankind disposes of intelligence as well as extelligence. Extelligence is the acquired knowledge to deal with matters not relating to its survival. A part of this development was due to the need of extelligence to eat. Mankind's progress in becoming a forceful species lagged behind because of his awareness to the fact that he could manipulate certain forces in nature to his advantage. A part of this manipulating process was accomplishing the skill to control and use fire. However, man also noticed that fire came from the heavens and clouds. These clouds formed part of the sky where the sun, moon and other stars are located.

With this argument, astronomy had to play a huge role in the development of mans culture, especially the religious aspect. His health, happiness, belief, future and wealth all derived from the gods that was found in the stars. This fascination and even religious fears was derived from the stars that even today is still apart of the science of cosmology. That is why even today, people are still motivated by the stars. Ironically enough, the other big motivation lies in man's lust for power and his war games to commit murder, to demolish other's property, and to dominate other members of its species to the point beyond that of slavery. The role of slave owners and slave drivers today is in the hands of the Mammonites. They use John and Jane Dow and all mankind that belongs to social grouping lower than they do. Mammonites are those that control every facet of the man on the street's life, should it be by job supply, political law enforcement, food and house supply, by dictating to the politicians in what manner and which laws should be applied and enforced. Mammonites are the bankers, the Wall street brokers, the Insurance firms, the drug lords, those that Motorcar manufacturing belongs to, the Shopping chain owners, the Oil barons and the Petrol companies but to name a few and all those and all other evil proprietors of wealth and monetary fortunes that force the law of the merchant onto the public. The merchant never has scruples, never has a conscience, adhere to one God which is Mammon has love for wealth alone and only has greed as a driving force. They that ask in all sincerity "What is wrong with having greed" but that question is on the lips of all other criminal elements in society. Mammonists on the other hand is the smaller and lesser counterpart that would pass his hungry brother and not help him, although he has more money than he needs, but he shares in the greed of the Mammonites.

To them, their love for money is far greater than their love for their fellow brothers and sisters and they will force millions to go hungry just so that they could have billions. This can be found in any social structure, be it capitalists, socialists, communists or kingdoms. They are the ones society looks up to while they are the one buying the politician to change law that bleed the population dry to enrich the those that already have everything anyway. Al Capone was nobody. Al Capone had one house in a neighbourhood of wealth. The others were more crooked and bigger gangsters that he was because they owned more than he was. They only pretended to be legal while they branded him as illegal. Who were the crooks that owned the other houses in the suburb that houses the rich? They are as big Mafioso because they pretend to be cream of the social structure while they would bleed those below him dry without mercy and always to the cream's advantage.

The fear that man experienced about forces, which was, according to him, inexplicable and therefore stronger than him, was considered by him to be of a godly nature and therefore only the wise amongst the wise could explain and philosophise about the nature of these forces. Common man never questioned the correctness of these layouts. However, man still prevailed in explaining to the best of his ability, the logic about his viewpoints and in so doing to guide the incredible forces that torched his fear. Today in retrospect, we consider these arguments laughable. Think how ridiculous it seems to regard the sun as a god on a chariot of fire that patrols the sky on a daily basis. Today it seems ridiculous to regard the earth as being flat, or to see a face of a goddess on the surface of the moon.

However, the tendency to except these super intellectual's arguments remains a custom until this day, no matter how silly it seems to be. Man is still upholding the ancient culture that in a case where it cannot be understood, the idea must be correct and therefore the ordinary man would inevitably be too stupid to follow. In addition, to this day, the "SUPER- EDUCATED- MASTER- OF- FACT" among us still misuse this phenomenon in common practice. Think about the silliness of Einstein's single dimension theory. How ridiculous is such a notion. Once we accept the earth to be round, Einstein came along and invented a flat universe!

In a hundred or two hundred years from now, it would be our generation's turn to be considered short sited and backwards because we accepted these ridiculous arguments. Thus, no matter how dynamic our visions of the cosmos seem to be, we shall still be regarded as non-intellectual and stupid by generations to come.

In writing this book, I too attempt to deliver a contribution to clarify a certain line of thought that is unclear and to give an explanatory value to it. You, the reader, will evaluate the acceptability of my reasoning and you will remain the only evaluator that will approve or reject my work.

My viewpoints are not the consequence of a big literacy, but rather due to a lack thereof. Because I shall entrust you as the reader with my thoughts, I shall have to introduce myself in a brief manner. Relatively spoken I can be considered as stupid. I do not try to sell myself short, but in accepting this fact, I was able to use it to my great advantage.

First, I have to qualify my statement that I am stupid. All people know what they know. However, they do not know what they do not know. For us mortals, the sum total of what we know, is enormously big. That comprises the total amount of our total human existence and our accumulated knowledge gained over a lifetime of labour. On the other hand, we regard the part that we do not know, as so insignificant small, that it bears a value of nothing. Because we do not know how much we do not know, we cannot evaluate the sum total of that.

People always concentrate on the part that they know and therefore realize how intelligent they are. In this lies the accumulation of their absolute arrogance. With this arrogance the part that they do not know, becomes even more insignificant. The normal procedure of man is that he will concentrate on the part that is of value, disregarding the worthless part that is of no value.

In my case, I had to concentrate on the part that I did not know, because of the lack of formal education and therefore not knowing how much I knew. This brought about that I always had to regard the part that I knew as being insignificant small. I had no formal examinations in testing how much I knew and thereby evaluating my field of knowledge. In the absence of examinations, I had to disregard the amount that I know and always had to consider myself as being stupid. This brought about that I had to remain humble, because I was untested and stupid. This book is the consequence not of my enormous intellect, but rather as the result of my stupidity. I always had to fight my ignorance and had to seek answers to my own questions because I was too stupid to accept the official answers given by the educated..

At school, I was a rebel. Because I was so stupid, I refused to accept facts without testing the reasons behind the answers and just because the teacher said so. An intellectual person would have accepted those given facts without causing him self all the inconveniences. I always insisted on outlined explanations by my teachers. I bluntly refused to "learn" anything. My point of view was that if I understood a subject brought about by a good explanation, studying was unnecessary. If the teacher could not explain the subject in depth, then I disregarded him and his subject and treated him with disrespect. I reasoned that the blind could not lead the blind.

This of course enraged the teachers and they tried to break my resistance with corporal punishment, which I deserved. This went into a spiral, where I got more rebellious and they had even more reason to apply the cane. In the end, I was the one that got the short side of the stick due to my rebellious stupidity, and brought about that I had to teach myself all that I know. I tell you this to point out that since my earliest days as a child I could never conform only because my superior said a certain thing and I was supposed to accept it. Sometimes (I guess) I was wrong and they were correct but only convincing me with intellectual arguments could allow me to see the other side. I was ultimately the one that paid the penalty for my behaviour because the road I took was tough. Some things I had to go through because of my stubbornness I do net even wish on the devil himself.

As I said, I am plainly stupid. But in saying that, I believe I am just your everyday person and if I could write this book, being as stupid as I am, any high school pupil can read and appreciate this book.

Because I do not have any noticeable academic background, I believe that these "SUPER- EDUCATED- MASTERS- OF- FACT" Academics will try to shoot the information down in flames, (if there are any that even would read it). The hostility I received so far from these Super Intellectuals did not surprise me in the least. However, I spent 21 years of my life to come to the conclusion that I share with you in this book and like every human being, I would want to defend my work against the onslaught of these sublime intellectuals.

The first few chapters are everyday common sense, but by regarding it, it will enable you to grasp my line of reasoning, which you will need in understanding the last few chapters. If one does not read the simple chapters, the terminology used in the last few chapters might seem somewhat incomprehensible but that is only because of new terminology that compelled me to introduce new terminology is brought in, in order allow he reader miss conceptions used and the new definitions I introduce.

This book must comply with a commercial value, to pay for the publishing costs and to introduce my work to the broadest range of reader's possible. That is the only way I believe I can force the academics to take note of my work. People associate cosmology with Einstein and his complicated brilliance and brainpower. That complication is only because Einstein told half the story. When I tell the other half in this book, you will see that it is not half as complicated as it seems. Every person with a normal mind will find all those unbelievable complicated statements that Einstein made, is in fact rather simple, when told in its full content.

On the other hand, this book must comply with certain technical facts to stop those "SUPER- EDUCATED- MASTER- OF- FACT" super intellectuals from blowing the statements that I make away with a few words. This work comprises of hundreds of new thoughts on cosmology, laws and processes in nature that was never noticed before and arguments that is now seeing the first light of day. Many of these arguments might be old statements that is purposely withheld by scientists, to the general public, because in admitting to the follies that exist in modern science, they then have to admit that their scientific layouts are in the least, foolish. As I said, the rejection received up to now, will be but a drop in the ocean in comparison to what I expect when this book is published in English. In the past, the only rejection on their part was because of the lack of my education, and therefore they refused to listen to my arguments. Therefore, I belabour this point concerning my education because if any person feels a need to reject my work based on the lack in education on my part, do it from the start, but if you do so, it will be to your own disadvantage.

I realize I do not beat around the bush, when confronting certain statements by some ingenious jokers that cannot be taken seriously. For that reason many of the academics would be sensitive to my work, but if they feel the need to make foolish arguments on international T.V. and in books, they must prepare themselves to be made the fools they are. These same intellectual gurus go to extreme detail in their own publications how the Roman Catholic Church denied science freedom of speech and prosecuted scientists in the dark Middle Ages. In that time, such people were prosecuted.

The powers that control the media today is far more powerful and much more methodical than what the Roman Catholic Church was then, and the modern day media's inquisitions are far less merciful and more subtle. If there is a certain school of thought, whom the modern media does not want to propagate, it will be killed by silencing its publication. If not for a medium like the Internet, this would have been the lot of this book as well.

This book will serve as a modern test in press freedom to see how science currently will respond to a new school of thought when their beliefs are ostracized. This book will put the shoe that was 500 years ago on the foot of the Roman Catholic Church, right back on the foot of science and the Newton apostles.

It may seem that I have a hate campaign against the intellectuals of the day. It is not true. However, I refuse to believe that with all their geniuses combined, they are unable to see the facts I have seen. I may be wrong, but there seems to be a sinister motive in the published work of modern science in that they promote atheism and use every chance to degenerate Christian belief. If a "mister nobody" as myself can see these facts so clearly, surely they have to see it too. Why do they then keep silent about it? They are the ones with almost unlimited IQ's, not me!

You might find the first four chapters to be simplistic and uncomplicated but if you do not get a well grasped understanding of the introduction of my theory, the complexity in the last four chapters will then seem impossible to follow.

The contents of this book is not aimed to relax and entertain, but the possible enrichment of your comprehension to the layout of God's creation will richly compensate for the effort, especially in the last four chapters. Thus, if you may find the first few chapters below your mental capacity and development, I ask you to bare with me, it would be worth your while when you reach the last four chapters and the truth starts to dawn on you. However, the last four chapters would seem complicated if the golden thread were not drawn right through to the end.

The main contents of this book, is as far as my knowledge goes, never been written or spoken by any person dead or alive and should be fresh to all.

If you do not approve of the very simplistic mathematical calculations, please feel free to ignore it. As I have shown in the prologue, it is merely put in to prove a point. If the simplest calculations on gravity are ridiculous, there is no applicable scientific applicability on reality. I did use it to point out how illogic certain scientific arguments are, but is will not enhance the explanation of this theme in any way what so ever. It is merely placed in this book to silence some of the "SUPER- EDUCATED- MASTER- OF- FACT" intellectuals.

As a background sketch to how I was motivated in writing this book, I must share the incident with you, being my reader. Due to my interest in the science of cosmology, I get asked certain questions from time to time to explain these known and accepted theories, principals and definitions propagated by the "SUPER- EDUCATED- MASTER- OF- FACT" academics.

As my personal studies progressed it became increasingly more difficult to explain certain accepted theories and concepts promoted by scientists and more even, to agree with these hypothetical mumble jumble and fairy tales. How does one defend certain accepted ideas that science promotes, but is faced on contra dictionary of how you relate to these facts? How does one explain your own concepts when it differs completely or does even vaguely been accommodated in the science of the day. I would have been able to ignore these conflicting feelings, if I did not know there are millions of John Dows out there who, as I do, did not appreciate (like me) or understand (and therefore do not accept) this misleading information.

To consider yourself part of the intellectual cosmology know how elite, certain recommended directions for use should be meticulously followed when confronted with the unexplainable concepts that is being promoted by the intellectual of the day. Applying these evading methods is very unacceptable to me. However, these directions are being used by the utmost intelligential on ignorant persons. When being confronted by a question you do not know the answer to, throw a mind boggler and complete mesmerizing question back at him, with the knowledge that nobody on earth knows the answer to that question. The questioner would find himself so bewildered that by the time, he recovers his senses, he would find himself still without an answer but he then would be out of a chance to insist on an answer. That will keep him ignorant and well in his place. On other occasions, other methods can be used with similar results.

First, if being asked an unanswerable question, congratulate the questioner on his brilliance and well thought question. Share the brilliance and well thought question. Share the brilliance of the question with the whole audience. Let the audience comprehend how brilliant the question really is and let the audience applaud his brilliance. Then abruptly ignore the question by going on to the next question. The first questioner will be so pleased with his own brilliance, he would never insist on an answer! However, beware; these methods can only be used in extreme cases and definitely not too often. The actual rudimentary way of dealing with the problem should be as follows.

(i) When nobody understands a certain concept give the questioner an impressed but unmistakable superior smile.

(ii) With a tone dripping with sympathy, you should use a stance of high and mighty superiority and let the questioner realize that you sympathize with his intellectual shortcomings.

(iii) Let the person very well realize (still with dripping sympathy) that if he was blessed with your intellectual insight and capacity, these facts would seem trivial, as it does in our own case, and you do understand his mental shortcoming, but he has to accept his inevitable weak minded position.

(iv) When the questioner's reason becomes far to logic suppress with your own ignorance of these illogic matters, like gravitation and electromagnetism, time and the black hole, immediately and without any further hesitation and with all the haste you can muster, refer to rule number one as stated above. Do not delay another instant. I shall damage your personal reputation to a point of no return.

I know all these methods of question evading by those super blessed intellectuals because it was used on me so many times by some of the most renowned geniuses. It was because of this that I started my personal search some 33 years ago. I had no intentions ever to put it down in published writing, because the motivation lacked on my part. For the past twelve years I kept myself occupied by trying to introduce my findings while not finding an audience very interested in what I have to say.

One night the owner of the local, Wimpy, Johan, asked me what space was. After trying to explain by starting with the atom, I saw that I was making very little to no progress. Then I decided to put these explanations down on paper so that he would be able to read it in his own time. Before long I realized that I could only explain this by means of a book, because in order to understand one explanation, I had to explain the facts that leads and follows that explanation. Now I find myself more than a decade years later, still trying to bring across my point where I wrote the book in Afrikaans, but has to translate it to English because of the weak market for such books in Afrikaans and for what it is worth here it is. *"Johan hier is jou antwoord in Engels!"*

The problem is those that should be interested in what I write are not interested because they are those that think they know more than God and that is not meant to be blasphemous because I show later on that those "SUPER- EDUCATED- MASTERS- OF- FACT" would rather have the cosmos change and start to contract in stead of expand as it does than admit Newton had everything wrong all along. Then we have the other lot that think physics are above them and they don't understand physics all the while it is that the physics they can't understand is so crooked not even Newton understood his own physics.

Between these two options of prospective readers I have not yet found that big understanding about the message I try to convey. I hope this effort will be simple enough so that everyone not connected to physics will show interest and understand what is wrong with the physics they don't understand and those that do understand physics will see how big fools they are too understand the physics Newton couldn't understand.

Those that do not understand physics will see why they were brilliant enough not to understand physics because it is one big hoax and those that do understand physics I hope will feel as stupid as they are because they think they understand the physics that is a complete joke.

SUBLIMATION; The Newtonian Mythology

The purpose of this book is not to echo the **Newtonian** version of Greek Mythology.
The Greek Mythology had aimed to bring, what they considered religion to be, in line with their perspective on what they considered science, cosmology and astronomy had to be. In Newton, a change a change came about in the contents of the mixture, but not the ingredients as such. The Anglo American Mythology now preaches a religion called atheism, which, as were in the case of the Greeks, based on what they perceive science and scientific facts to be. If you find a desire to run along with these myths of star travel, speeding through the Universe at the speed of light, encountering alien societies and indulge yourself in such modern mythology, please do not read this book. In this book, the magic spells that Newton named gravity, and which Einstein took to a single dimension fantasy, is discarded in dismay once the Anglo American Mythology is replaced by factual truth, the creation according to the Bible becomes a detailed analysis of the actual creation. When Newton's lies are replaced by the real functioning of the universe, the author of the Bible's firs book has such a precise recollection of events, that it put all scientific facts, being broadcasted at present to pitiful shame. That includes the ideas of modern atheistic Anglo American science cult. This book does not repeat all the traditional nonsense, but explain in detail, how the cosmic year structure works, in such detail that children can understand and accept it.

Civilization throughout the ages always used Cosmic Science, but especially, by the Roman Catholic Church, to prove the importance of the earth as the centre of the universe. That brought about that the sun shone on the earth, which was the centre of the universe. Since the earth was made for man and was the centre of the universe, God made everything with mankind in mind. Whereas the Roman Catholic Church was the only representation of God on earth saw they represent God and all His Powers on earth. It meant that the Pope was God on earth (being the head of the church of God on earth) and the Catholic hierarchy was the most elite and privileged on earth. That was precisely how the church considered them and how they could conduct their teachings.

Right at the top was the Pope who ruled the earth and as the earth was the centre of the universe, the Pope for that matter, ruled the whole universe. God and the saints ruled the heavens and the Pope with the Cardinals ruled the creation of God. This brought about the sublime picture that suited every person, which was considered anybody. He whom the Pope blessed was blessed and he, whom the Pope damned, was doomed. Everybody op importance could buy the Pope's blessing and could die in reinsurance that his life was ultimately saved.

By the middle of the 15th century, some unimportant persons with no real social standing came along and disagreed with this ultimate and universal accepted hierarchy. They stated that the earth was not the centre of the universe, which meant that the Pope and his cardinals were not in control of the universe. This new perspective on the Universe and the chain of command was very unacceptable to anybody of social standing. This (then considered) blasphemy was to be killed in its very infancy, even before birth. However, as with everything else, the truth eventually prevailed and certain intellectuals took notice of statements by Galileo and Kepler. The reaction that followed can be regarded as one of the most important historical events in the route that man's civilization took in forming modern man.

In this book, we put the work of the two giants, Galileo and Kepler, under the magnifying glass. In contras to popular belief, there is an astonishing difference between the work of Galileo and Kepler and modern science. Modern science is based on the findings of the father of modern science, which I consider Anglo-American mythology. The father of Anglo-American mythology is none other than Isaac Newton in person. Galileo and Kepler had the truth of the Universe unlocked when Isaac Newton came along and raped their findings.

First, let us consider Galileo's work.

When a pendulum swings, the pendulum's rhythm remains while the stroke tarnish. All big clocks are based on this working principle and can therefore keep time mechanically. Time has been measured by means of this method for about half a millennium.

When two structures of different mass is dropped at an equal distance and time, the two structures will hit the earth at precisely the same time, as long as the wind resistance is equal to the two bodies. This experiment was the first that was done on the surface of them moon on July 1969 and billions of T.V. spectators bared witness to the outcome of this experiment. A hammer and a feather were dropped and they hit the surface of the moon on precisely the same instant.

Kepler, on the other hand proved that the earth and all other planets rotate around the sun in an elliptic orbit. All three these findings never mentioned any force called gravity.

At this stage, science used precisely the correct argument. The findings were noted and it could afterwards be checked for the same results repeatedly.

Afterwards an English genius by the name of Isaac Newton noticed an apple falling from a tree. This prompted him to calculate a force that existed between the apple and the earth, in which the matter of the apple and the matter of the earth was drawn by a force he named gravity. He never took into consideration any of the findings of the previous two giants, although he praised them for their work. Newton went and calculated the existence of a force that existed between the above-mentioned bodies. At this very point, science took a wrong direction. The force that Newton calculated is a secondary function to the primary condition that holds matter in place throughout the universe.

By using the findings of an even bigger genius, Tycho Brahe, Johannes Kepler proved that the radius of the planet, taken from the sun to the planet and is measured in astronomical units, is equal to the square of the rotation period. Kepler said that there exists a relation between two bodies where $T^2 = R^3$. He never mentioned a force. Galileo's findings proved the same as Kepler's, that there is a ratio between two bodies, not a force.

The second statement of Galileo can compared to driftwood on water. If two pieces of wood which is different in size and weight floats on water, and both pieces of wood is subjected to the same force value in the stream, both pieces of wood float at equal force, no matter what the difference in size and weight is that comes into play. This would be caused by the difference in the drag resistance on the different surface area of the two wooden bodies. However, the difference in weight that comes into play, at this point is only due to the drag that the water experiences, not the actual weight.

If this were compared to, the findings of Galileo one would find that this is the precise method how matter moves towards the earth. Galileo made no mention of a force between the two bodies. Then Newton came along and published his mathematical findings, which is totally out of line with Galileo's findings and ever since then no person ever gave the actual findings of Galileo and Kepler a second thought. There can be no force such as gravity, electromagnetism, strong and weak forces, or nuclear energy. These so called forces are part of precisely the same value that is in relation and exist between space and time.

You, the reader may ask yourself why is there such importance in these findings as to know the correct way that gravity actually works? All calculations have already been made; mankind already possessed the knowledge, expertise and willpower to visit out of this world's Tara novas and even to colonize them.

All knowledge about physics and astrophysics has been studied, formulated and tested! I will reply to this question by asking two other questions.

A certain man drives his car down a lonely road. After a while, the car comes to a standstill. He knows that a car needs fuel to run. He takes a 25-liter can and walks 10 km to a filling station to buy fuel for the car. He carries the 25-liter fuel 10 km back to the car and puts it in. After trying for 10 minutes, the car still would not start. He takes the 25-liter can and walks 15 km to another fuel station to buy some more fuel. After walking all the way back and filling the tank with the petrol, the car still refuses to start. He then takes the 25-liter can and walks 25 km to yet another fuel station for fuel. After returning, he filled the tank yet again. Do you, as the reader think for one minute the car would start?

The American scientists spends 1 000 million dollars to pressurize four hydrogen atoms into the same space of one helium atom. The experiment seems unsuccessful because the helium unfolds back to the original four hydrogen atoms after four seconds. After that they use three 000 million dollars to pressurize four hydrogen atoms into the space that one helium atom occupies. After seven seconds the four hydrogen atoms depressurizes back to its original state. Now the American nuclear scientists use 10 000 million dollars to pressurize the four hydrogen atoms into the space of the one helium atom where the experiment lasts for 12 seconds. After 12 seconds the hydrogen atoms moves back to their original condition. Question 2 now is this: "Can you see any connection between the two examples I have put to you? "No pressure in the world can fuse four hydrogen atoms to one helium atom, even if Einstein said so! Those super brains and academics are blind with their own mathematical genius of mathematics and physics that they fail to see the most basic and elementary principals in science. What is tragic is that it does with the taxpayer's hard earned money!

Another rudimentary example is the so-called "falling star". The conventional theory that is propagated is that the dust speck burns to ashes because of the friction the particle has with the air it collides with in the atmosphere. This is the biggest mindless rubbish one can imagine. Just because the person that tells me this have six doctoral degrees, does not make it the gospel truth. Far from it... What actually happens is that when the grain of dust enters the earth's atmosphere, the time aspect changes and forces the dust particle into a different space occupation which then changes the heat value of the grain of dust. The dust grain is forced into such a smaller volume of space-time occupation that the heat it generates just burns it to ashes. Therefore the whole structure glows itself into nothing.

These "SUPER- EDUCATED- MASTER- OF- FACT" giants might use the most breathtaking mathematical formulae that can humanly be dreamt up, but if the principle, on which they have based their calculations on, is wrong, the whole exercise is fruitless.

In my book, I discard the "conventional" standpoints by following the unconventional principals, based, on the work of Galileo and Kepler. The basis for my theory has never before been propagated except for me, and therefore all arguments will be new and fresh.

I am no writer and even less a scientist. Furthermore, I do not pretend to be regarded as any of the above mentioned. I simply came to certain conclusions. When I shared my conclusions with other people, I had the very same reaction repeatedly. The reaction followed spontaneously without exception. Those who were prepared to listen to me knew even less about cosmology than myself and never understood a word I said. Those who know more than I do ignored me immediately when I said Newton and Einstein are wrong in their views about gravity.

Partly the writing of this book is to prove that there is little difference to science and the Bible, even if scientists and theologians try to make it their lifetime task to prove the other opposing side wrong. However, all their arguments are based on their individual agnostic belief in their self-righteousness and have nothing to do with the truth.

Free thought has always been a fact that all that is in influential positions proclaim to strive to but the minute the free thought differs from their concepts, it is the very first thing they crush as hard as they might. In this book, I strive to accomplish the very essence of free thought and try to lead the reader away from the brainwashing that all intellectuals try to force onto the public. Every person that is in any position is busy with their own sublimation that it renders them blind to the truth that is out there.

Man-In-Motion, Man-In-Mind, Man-In- Motive, Man Is Blind.

I wish to state here and now un-emphatically and categorically without any reservation of any kind there may or there may not be that it is totally against my personal religion as it is against the religion my Congregation upholds that I belong to in converting whom ever for whatever reason and never, never has this article or any other reference I may make any purpose in converting any body to my way of view about the spiritual in any way. It is not what you believe or not believe but how you live by your convictions and what ever your convictions may hold making you man or beast. My remarks about atheism are to show those practising atheism the foolishness of exclusion and to have an open mind because teachings of what ever nature has a positive and a negative connotation and the individual sets the standards. I am not intending or have any intentions in converting or changing any person's outlook on the spiritual side even in the slightest way imaginable. Me, living by my conviction to its full believe no bigger sin can there be than converting a person for that more than any other fact holds the highest epitome of sublimation there are. Secondly and in line with the first is my belief that the Bible has a base derived from ancient Egyptian teachings and I refer to that as I go along exchanging thought in this article. To this day we with all knowledge of splendour can still not understand how the Egyptians erected the colossal structures and the manner in which they did it. It is not realistic to consider a civilisation being that advanced in one area exclusively and with no other wisdom in other advances.

The hour in thinking has dawned where we humans must come to terms with the cosmos, with creation and with life. Mixing and matching was fine up to a point in the nineteen sixties where thinking about the cosmos and creation was a smart way to show superior intellect but being rite or being wrong was only a case of honour and pride that can hurt. Since the sixties life loss results from incorrect principles and maters got far more serious than it had since man had a first time ever look at the night sky with a conversation beginning from that. In another book as part of THE THESES I show briefly why I am of the opinion that man became human when he saw the funny silver dots in the night sky with a degree of admiration and recognition to the splendour of the unknown.

Now there is no longer only splendour in the unknown but a quest to find the unknown and gallant as they ever may be, it is fool heartiness to send brave men and women on search and not know what are there waiting on them. What dangers will establish the outcome of their fate? It is not philosophising for the pride that we should find evidence in distinction where distinction should be but crucial to man and to machine, machine because man's life interlinks with machine as it did at no other time in development history of man. A great philosopher I may not be and I will never be but thinking does not hurt and some advances may come from the weakest of thoughts when the thought try to unravel a thread running in thoughts. With this I too wish to connect in sharing thoughts of woven patterns as I see them and for what they ever may be worth.

I WISH TO TAKE NOTHING OUT OF THE UNIVERSE AND SHOW IT AS THE UNIVRSE IS: AN OVERLOWING CONTAINER FILLED TO THE BRIM WITHOUT THE SMALLEST FRACTION OF EMPTYNESS ANYWHERE.

As I stand on earth holding my first dimensional space-time displacement of our planet I can observe by using the second dimensional light source of the sun where my surroundings are made of three dimension atoms holding space-time in the forth dimension in time and space. I wish to move from point A to B and think in consideration about my planned action as my brain sends electric impulses to my muscles and that brings my muscles in mechanical motion. Arriving at point B I think to stop (not necessarily by thought or mental planning) and my brain stops sending electrical impulses to the muscle fibre concerned with the action in applying my body motion. At that point I come to a stop and my thoughts go to a rugby match played at Loftus a South African provincial rugby team's head quarters where a game is in progress at that specific moment. As the proverb goes: I am there in spirit and my spirit being at Loftus are then some 400 km. south of my body where my body is on my farm. The duration in time it took my thoughts to travel is beyond human measure

The very next instant my mind goes much farther back in time and space as I travel by mental motion to a game played in Christ Church New Zealand a week prior to the day in question. My thoughts took me not only half way around the world but out of the present time dimension. As I am standing in thought I see my next-door Neighbour (to us in South Africa your next door neighbour lives normally 30 km. from you) coming towards me and my thoughts return not only from New Zealand but also from a weak past to the present in the very current time span my body was occupying all the time while I find my body moving towards my neighbour without my actual realising of this moving motion. I use the second dimensional system in the wave to transmit sound by means of repositioning the three dimensional atoms between Old Neighbour and me in applying motion to the atoms between us by the fourth dimension of space in time to convey thought harboured in the fifth dimension to him being in the fourth dimension of space and time. He then uses the same system to convey a thought by massage to me. Please note that it is a thought from the fifth dimension that I convey with the applying of organs in the forth dimension through ordering electrons in the third dimension to control matter in my body placed in the fourth dimension of space in time. While my words carry towards him, he drops down like a log as a result of not fighting the first dimension called gravity. A thought from the fifth dimension prompts me to respond in the forth dimension by creating electrons in the third dimension while my body stands supported but restrained at the same time by gravity in the first dimension. The thought from the fifth dimension orders a response in all the other dimensions ordering my atoms in the third dimension to use the forth dimension of space in time to act by fighting gravity in the first dimension on what I see in light holding a place in the second dimension which is restraining my motion in the forth dimension.

My response comes from some emotion and as it is not part of my mental reasoning or thought pattern it is directly conveyed form the fifth dimension to respond. I feel his pulse and find no beat. His breathing stopped and my next action is to look into his eyes. There is a dullness in his eyes that was not present moments ago. Something went that was. My observation consists of thoughts relaying massages that is transmitted by my physical body in relation to my senses receiving and responding to electronic massage translations about Old Neighbour in the fact that he somehow relinquished all earthly responsibility and problems of an earthly nature to his next of kin that is now saddled and burdened with his last remains.

The heart shows has no beat indicated by the absence of a pulse. The longs lost all ability to provide oxygen for transmitting and burning food. His eyes became stony marbles. I communicated with him moments ago, but his ability to respond by hearing and speaking has gone absent. My thoughts and breathing, heartbeat and hearing are still there. I can speak to him but it is his ears that have gone deaf. I can squash air down his longs but he is unable to use it through his voice box. I can hit his chest with fury and support a heartbeat, but the blood will carry the oxygen but can no longer create heat to live.

The air I force down his throat still has the ability to produce sound because my shouting to him creates sound, but his ability to establish a method whereby he can create sound from the air I push down into his longs has gone away. His ears are still connected to his head and all the required tools equipping all previous aid that use to enable him are still there, all intact, but also gone forever.

All the biological organs needed for hearing and making sound is still unscathed in the right places not damaged in the least, but the use has gone. The electrons needed to translate whatever requirements enabling body function must still be in there somewhere, because I saw no discharge of any sorts flashing from his body.

Even by giving him electrons through an externally generated flow of electricity will not create any of the required but lost electron flow to generate life back in place. I may shock him till he hops around all over the place, but motion is denied for brain activity to function once more. His brain has gone empty, although it is full to the scull. His thoughts are no longer with us or with his body. It is no longer Old Neighbour lying there, but it is his remains. Even an atheist will tell you there is a difference in what is there on the ground and what were there moments before he dropped to the ground. No heat or electricity can revive what he lost. It is a body without life.

Minutes ago there was life to talk think and reason, discuss and argue, be angry or glad, but that, which now is lying on the earth has no more such ability. The source of energy giving life to the cadaver is no longer present. Science proved that energy cannot go lost but has to go from one form to another form. Energy can never destroy or vanish but has to replace form or attachment. The body is there, and it is holding all the organs and the organs has still got the required heat to perform because Old Neighbour has not gone but a few minutes ago and in the South African sun bodies do not go cold through lack of heat because we are use to temperatures of forty degrees Celsius and more.

The cadaver has all the essence to sustain life and if life was electricity, then I should be able to recharge him by connecting leads somewhere and call an ambulance. But supplying any form of current at any voltage rate will not bring back life once it has gone. You can heat him with a blowtorch while shocking him with a cow prodder (and does those things unleash electricity!) it will revive him as much harm him or do him bad or good. He has become apathy in every sense.

His lifeless body will never carry his mind anywhere again because although the brains are still there holding all the mass it had when Old Neighbour was still with us, the brain is thoughtless and that has taken Old Neighbour away from us. Our dearly has departed although his physical remains stayed with us to rot if we do not take care of the cadaver and the sooner the better for everybody involved. We that are part of the living now have to move Old Neighbour because he no longer has such abilities. Minutes ago he still had the abilities but from him went energy. It must be energy that he lost because all other necessities in for filling such duties he still has (that is if you consider his body as Him) But his body cannot be him because his body is there part of the fourth dimension in space and time securing all his abilities to function as a human but that abilities has gone vacant. All the effort he may muster will not allow a wink.

The only visible something he lost that makes him less of a human being than he was this morning when he woke from a nights sleep is the energy of motion. His body with all the parts still hold dimensions in the first the second the third and the fourth dimension, but clearly it is the fifth dimension that has gone absent. The cadaver is still part of every dimension excluding the dimension of life and life then has to be a dimension above and beyond that of the fourth dimension in space and time. The cadaver is at present what we refer to as being lifeless and dead. The generator or power source or dynamo or what ever you may consider it to be but that dynamics providing energy in sustaining motion has gone away never to come back.

Whatever any person may try to do the machine that gave drive to motion is no longer able to provide motion. All the wonders that the human body possess in motorised function is no longer in motorised function although it should be if it was only a matter of replacing the lost energy by providing an electrical shock or some fuel of some sorts. Nevertheless no fuel can get that motor running again therefore the energy lost is not a replaceable kind as in the case of ordinary heat from fossil fuels, food or electricity. The machine of human motion has gone for good. Surprisingly the problem of energy and life becomes far apart when logic replaces Newtonian atheism and illogic.

Shove a ton of coal down his thought and it will do him or you no good at all. Roast him with electricity and see how far that will convince him to return to life. Push a gallon of pure glucose into his veins and see what the reaction will be. Energy is not merely energy and once again Newton got every thing very wrong. With Newton's incorrectness all the sheepish atheists go about an echo one can hear for miles around, but all the echo is only echo after all with no substantiating individual thought about and amongst the lot of them atheists. When energy is not used it becomes latent or so does science proclaim in any case.

One cannot ever consider a rock rolled up a hill having the same latent energy because the rock needed the same life that has gone absent from Old Neighbour to role uphill in the first case. When inspected closer life is the energy keeping the human body running as a motor and by distinguishing life the motor stops. Something went latent and not vanished. Life was part of the fourth dimension up to the moment it went latent. It shared time in the body and space with the body thus it was part of the matter of the body it no longer uses. Without doubt is the fact that life was the indisputable source of energy driving the body through the fourth dimension? Where the space-time sharing then ends, life cannot end because it is the functions of the body ending and not the energy driving the body while inside the body when it gave the body a function of movement the body had an ability that no other cluster of atoms enjoys in one construction in the universe. It gave the body the means to displace space-time not only by gravity and motion as all other structures have but it gave the body a means of changing the space the body occupies in time that the body occupies. No mountain can move a little in the morning to avoid the blistering sun and shift to another place at night to escape a blistering cold wind.

The human body including all life on earth can shift position as to suit the needs and requirements of space enjoyment in time duration. This means is very exceptional as nothing ells known to man in the cosmos can achieve such motion by pure will power. A plant may not be able to run to a better position but when in competition for sunlight it can try to outgrow its neighbouring plants and claim a larger share of the available heat the sun has to offer. That effort is completely out of the domain of any rock. A plant can grow its seed in such a way as to ensure distribution and gain advantage over the spread of its space it holds on earth as territory. There is no chance that a puddle of mud can run after water to keep wet. Life can manipulate the space-time it holds to its advantage in the sense of bettering its chances on survival as well as its species chances on relocation.

That is the overall advantage life holds and is not merely an energy that does some work in relation to the growth in the universe. Try and measure the time a mountain holds space and compare that to any one form that life holds measuring from birth to death being on this planet. Then after getting an unbelievable answer a person can appreciate that life is the energy and without life the structure becomes the equal to what a mountain is from the onset of the lava flow.

Life is the manipulation of space-time and the higher the degree of advance is, the more life can manipulate space-time. An aircraft flying may be as dead as the next mountain is, but through the aircraft, man as the ultimate form of life can manipulate space-time far outside the reach of lesser species. I do not wish to start comparing life as being advance or more advance so I leave my argument at that as far as life development goes for now. In the very beginning I stated that through the way the mind travels, it has to hold a higher position than what the body holds because I showed how easily I could travel around and even half way around the globe in no time at all. Sure I was not there in person, but my thoughts conveyed some understanding of what was happening on other places outside my range of vision.

The mind sets a norm that the body can follow or not follow but the body never sets a norm that the mind cannot follow. All sells even those holding life has an electron a neutron and a proton and very deep within the very deep within next to the truly unknown is a structure that holds position in relation to singularity. When a sell holds life it is different from a sell not holding life although when the sell not holding life still constitutes of the same composition it had when holding life, something changed, something is different.

A life-carrying sell not carrying life has gangrene a most deadly disease that kills as none other. A sell absorbing heat normally is showing growth whereas a sell in abnormal heat intake is cancerous and again is deadly. I can go on and on about this but it is apparent that as soon as life looses control over heat the stabilising factor or thing go abnormally wrong and such conditions can, may and will lead to the vacancy of life occupying the body more permanently. To understand the way I wish to direct the argument please allow me to indicate how I see the normal as we will find in the cosmos in life carrying and non-life carrying matter.

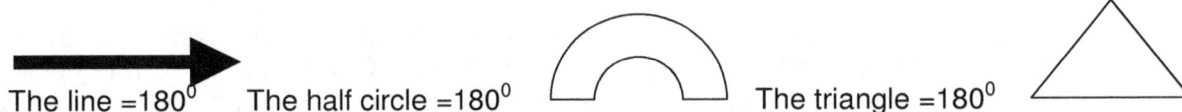

The line =180^0 The half circle =180^0 The triangle =180^0

To begin to understand human nature we must first understand physics and to begin to understand physics we have to understand mathematics. There is a reason why mathematics is the way it is. Any student that has a Professor that the student worships for his or her mathematical brightness and think the Professor is in some way a mathematical God-like on earth by the cleverness the Professor has in interpreting mathematics. Go on test your Professor…Tel him or her but make sure you ask the Professor to tell you why would a straight line and a half circle and a triangle be the same value. How can a line be the same as a half circle and a triangle? When I came to the answer of this, I found hiding behind this answer is the manner in which the Universe formed. That is how the Universe began, but it took me a very long time to figure that out.

Since almost before serious recorded history dating became scientific principle mathematicians knew that the straight line holds 180^0 degrees matching the half circle as well as the triangle. But never have I read any definition about this phenomenon and how it comes about or what may cause such odd connection.

Heat occupying space has the cube that can apply r, as a straight line bringing about the cube with all its other names that may find attachment to specific form but nevertheless still remains only a six-sided cube with angles changing in some cases.

Creation is not there to serve life. Creation is not made for humans to enjoy and to multiply and conquer everything in the universe that we see. We with life had to adapt what is in the Universe to make the Universe work in terms of what we need. The Universe does not have to apply what we have because the Universe was in place long before life became a thought in terms of the Universe. What we find in physics will apply to life and in detail. Life is a designated factor of physics within the universe. The thought of life is not as simple as the simple minded Newtonian atheists think it is because the simple minded Newtonian atheist can only think in terms of resolving complicated issues by simplifying extremely complicated matters down to what the simpleton with his mathematics can formulate and prove. Let the Newtonian atheist prove by using physics we can think. I can and I do that in other books. I wish to keep this book simple and therefore I am not going to get complicated and going into that type of detail get quite complicated. Then if our simple-minded mathematician calling himself a Newtonian can't prove mathematically how a thought is created in the mind his physics come short on a gross scale. Don't begin with electrons flowing through the brain because going that direction you then presume that a generator generating electric current is one of the best thinkers we can find…and I will not be surprised to find the Newtonians think this way. They are of the opinion that computers can replace human life and that proves how inconsequential they argue about facts that are way beyond what they ever can understand. Life is partly the ability to generate electricity but that if what life can accomplish, not what life is. We have to look at physics and not Newtonian half-witted rubbish to see how the human physic works, and by establishing science we can see how life controls the body to act on behalf of the mind.

Let's take how physics work and see how that pans out.

In the very centre of the sphere the form of the sphere dictates that the shape will relinquish space as the line run from the outside towards the very centre. With this natural state of affairs the sphere are naturally inclined to dismiss all space that it can form in the form as the sphere holds space inside and the form will finally be without dimension. All that I attribute to the line shrinking by reducing actually takes pace in every sphere as the diameter reduces to the centre. In the centre where the radius line goes single the form relinquish the three dimensional form it has inside. Being without dimension in the very centre means that at a point in the extreme centre of all spheres there are a point that holds singularity because this point with no space has a mathematical position although it is invisible since there is no sides to such a point to give that point any dimensions.

The shape of the sphere is calculated by using the formula $4\Pi (r^3) / 3$. By reducing r to a point where r is r^0 singularity steps in because only the form remains as Π. Going even further we find that there then comes a point where Π goes singular Π^0. At that point absolute singularity is present but so is absolute gravity present at that point. When holding the strength of the shape of the sphere in mind as well as taking into account that all cosmos objects of importance is in the form of planets or stars and they are all in the form of a sphere, we therefore may contemplate that it is where gravity originate. We now only have to find the reason why gravity will hold a base in a space less ness as Einstein predicted. It is clear to be seen that gravity is in the centre of the sphere controlling from the centre everything that is outside the space less centre.

We can reason with confidence that gravity is the strongest where space is the least. We can further reason that it is gravity that is holding the sphere in true form and since the sphere allow gravity the best working opportunity, gravity can form the sphere in as strong a shape and form as the sphere seems to have. From every point on the surface of the sphere is where that point connects with the other side of the surface of the sphere by a line that runs through the space less ness of such a centre of the sphere. Such a line also connect by an angle of 180^0 as well as 90^0 to six other lines running from top to bottom, right to left, and back to front, where all join and cross in the centre of the sphere. There are therefore six lines crossing and connecting by a centre from any given point on the surface of the sphere. Such points connects in total six surface points on each side of the sphere while they all support one another through the space less centre. In that absolute space less ness in the centre holding singularity we find gravity supporting and controlling all space within the sphere as well as space connected to the sphere. That is where gravity control and guide the space, which falls in the parameters as well as under the influence of the form of the sphere. In the gravity centre space goes singular meaning space becomes space less or flat.

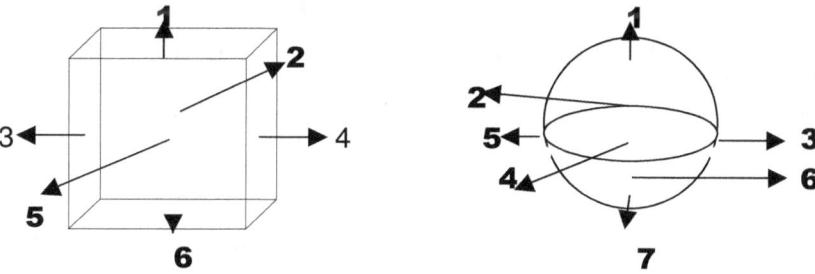

It is from the layout that the sphere uses as natural form that we are able to locate singularity. In the case of the sphere the material naturally reduces by measure of the radius becoming smaller to a point where the radius is r^0. At that point the line that will form the radius has gone single dimensional r^0 and that is equal to 1^0, which is singularity.

Also it is true that the entire form that is the sphere is controlled from a centre within the sphere. That centre holds the sphere in form and shape. Therefore the strong form is dictated from that space fewer centres where there is no space and no form left. The natural inclining is in the form of the sphere. It is part of the roundness that the overall shape of the sphere represents and this structural strength is carrying down to the very centre. Because the circle is forever reducing that reducing which is inherently part of the form of the sphere becomes a tool in distorting of space in the sphere and is eventually removing all forms of space from within the centre of the sphere. The very centre ends up as having no space because of the reducing that continuous down to become the space less inner centre. The all roundness is the ingredient that forms the backbone of the absolute strength that the sphere has and that is the component that the sphere is so famous for. The form the sphere has allows the sphere to have a control that is coming from the centre deep inside the sphere where the space vanishes and being without space seems to keep the entire structure rigged. From the centre the sphere shape shows strength that the shape as tough as it is. How does it work in its most basic analyses?

This spot I just described and pointed out has no place in the Universe and yet it is what controls the Universe however, the simpleminded Newtonian got this spot down to **"Nothing"** albeit the most crucial point in the entire Universe.

Where space comes into contact with the sphere the cube loses one of the six dimensions it has to the more dominating seven dimension of the sphere whereby the seven dimension in equilibrium will dominate the six dimension loosely connected bringing about that the cube then has 5 sides to the seven of the cube. This means that in the cube the "bottom falls out" and without a "bottom" to support objects they fall to earth. Remember that a body "floats" in space, but at one specific point it starts to "fall" to the earth. That is gravity and it is a dimension change much more than any force.

The spinning of Π^0 around the centre Π^0 establishes Π and Π is what produces the form gravity has. Still it is the relation or relevancy there is between the centre Π^0 and the spinning Π^0 that gives status to the form that Π represents. In out Universe we are accustomed to and are familiar to the rules we want to place seven points holding singularity to the centre holding singularity in a relation of $7/10 \; \Pi^6 / 6 = 112$. In that Universe everything less that a duplication ability to the value of 112 protons fit but only atoms to a maximum of 112 protons fit.

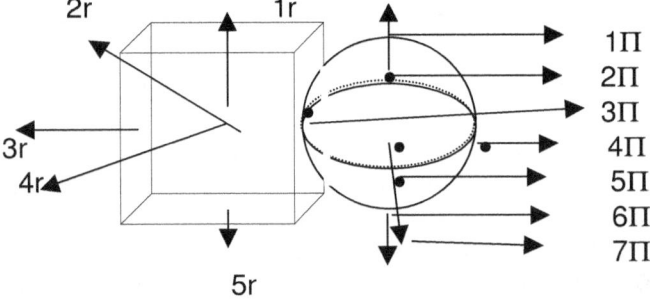

From such a point every other point will be opposing any other point not pointing in the direction to which the first point is pointing, whereby it extends the direction it holds. No matter what the point is or where the point leads, such a point holding a specific direction will be unique in the direction it is rotating because at that or any other specific point wherever, it will be directing not in the direction it spins but in the direction flowing from the centre point outwards.

All atoms are a minute form of a coming black Hole and viewed in the structure composition it is clear why I say this. On the outside there is heat trying to get inside the atom where the heat is needed. On the inside of the atom there is a need for heat and the inside is in constant regulation of the heat flow as to keep stability. In understanding the dynamics of physics we must understand the cosmos where the process begins and where the process ends.

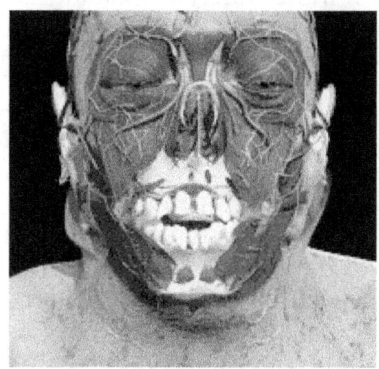

This is not the person but it is the tools life uses to enable the person to manipulate space-time in the interest of life. This is the body function but the body function adheres to the mind controlling the body. The body works by the mind establishing thought and the thought creates electricity and then with the electricity being a messenger, just as this computing machine uses electricity to convey what I have in mind through guiding electrons work is done firstly by telling the body to hit the keys on the keyboard and from there electrical signals send the commands to the computer to do whatever is necessary. The computer is not life but is only extending the ability of life. The body is not life but it is extending the ability of life. Life is in the thought and the though controls the body but is not part of the body.

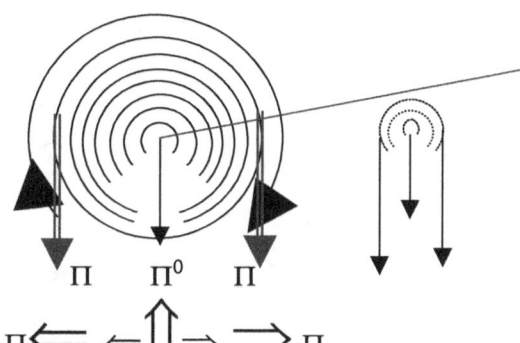

Pinpoint positioning of singularity I with Π positioning space to either si forming the border set by singularity.

The atom holds a very unique position in that it links three dimensions to a forth dimension and this part is where I came to understand Einstein's thinking but a with Newton I could not accept Einstein's thinking. Only bringing in religion could I get further about the formulation of singularity because in that I found what connected the universe whereas Einstein left space as space and tried to link time to space as an additional factor. That would be the same as not linking life to the body or exclude life from the body while trying to argue about life being part of the cosmos, as Newtonians seem to do.

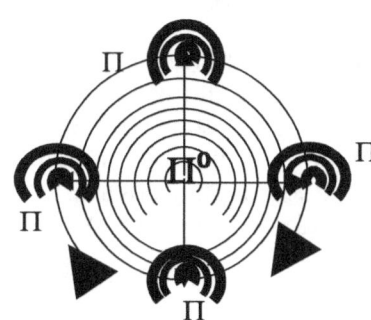

When singularity expanded for the first time ever and when heat from cold bringing about the Universe forming 1^0 to 1^1 from Π^0 relevancy was born and that relevancy grew into what we now ha a Universe

Gravity in the centre formed time Π^2 by dismissing while the fou positions started the cosmic trend of duplicating.

With every one of the four points taking form to the value of I measure of $\Pi/2$ each brought about the Roche value of Π^2 relation to the developing centre. One has to remember that th of today takes on the characteristics of the form of that era.

If you go down any spinning object towards the centre there is a spot in the very centre that is not part of the cosmos because that spot holds no space nor does that spot represent the Universe in any way. Yet the spot is present and the spot controls all space surrounding the spot. If any reader wish to know more about the spot I suggest you must read other books where I go much deeper intro the complexity of this spot.

The centre changes motion to gravity by diverting the straight line to an immediate circle. By tracing the line back to where the circle is no more a straight line will uncover singularity plus one dimension. However, the entire centre forming singularity is still locatable within the Universe we have.

Reducing the radius r from all angles possible throughout the circle will bring about that all possible direction will eventually land on the very same spot with no more dividing possible. Yet zero cannot be a factor since the sides still hold value. In as much as holding all the value there can rise from such a spot. This is arriving at a point where more reducing will land the one side on the opposite side of the line but it will not bring about zero in the equation

Only by understanding this concept can we go on and can we see that life as a factor is not part of the cosmos as singularity in the spot is not part of the cosmos.

r /2 By dividing the radius r by the half of the value that then reduces r to a point where the left edge of the line reducing will be at the very same place the right hand edge of the line that is reducing will be. At one point the spots that formed the two ends of the line will be at the same spot where the original centre between the two points were. The two points would have moved evenly towards and in the direction of the centre by reducing all the space on both sides of the centre. By moving towards the centre they will at some point have to reach such a centre point notwithstanding cultural concepts favouring nothing to be filling that spot. Reaching that centre point will land all the sides on the same side and because of the presence of all possible sides such presence of all possible sides removes nothing out of any further possibility.

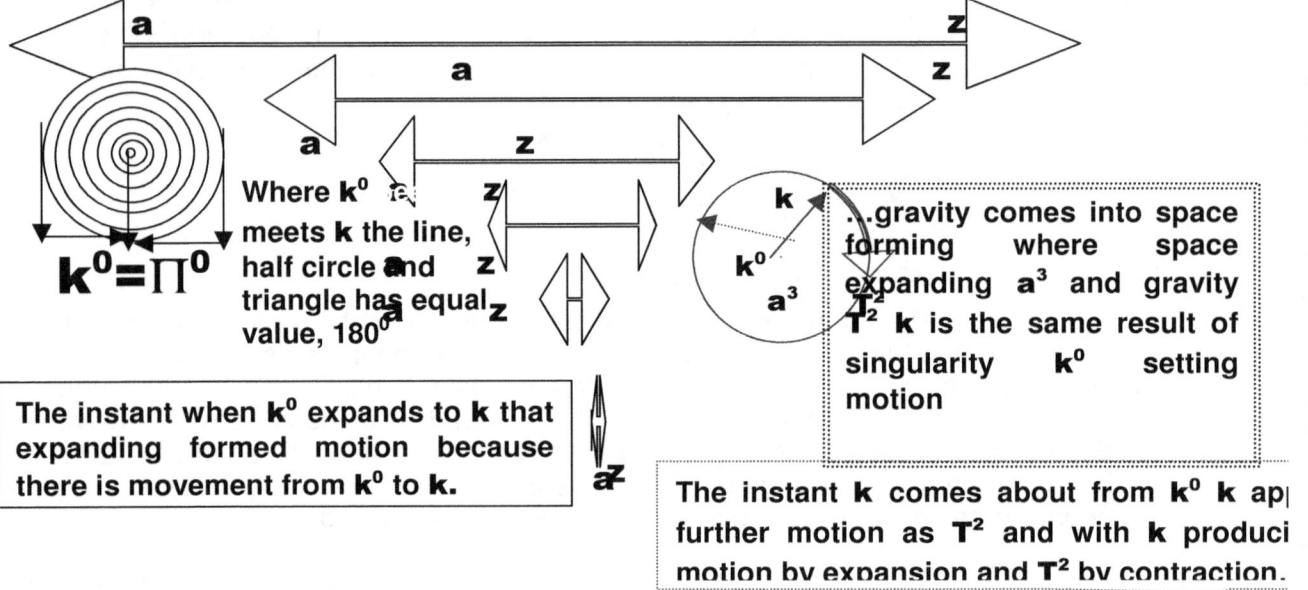

In the sketch I made, shows below each of the lines reducing there is a space left open

This occurs in all atoms through out the cosmos with no exception on the rule. But life-carrying atoms in carbon $_6$ commits life as an additional supplement to the atom as life can become absent from the atom leaving still in the normal range of a cosmic structure. In the past number of pages I brought reason to those of reason that there are more to the body supplemented by the presence of life than merely carbon fibre. It runs much deeper than physics can intrepid. As far as pure physics go, nothing changes when life goes absent and yet everything alters when life abandons the atom.

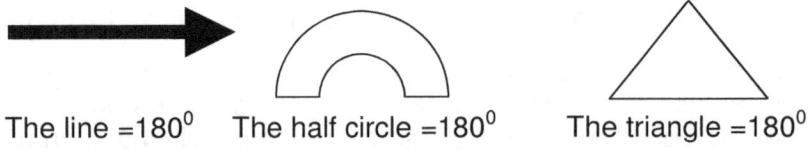

The line =180^0 The half circle =180^0 The triangle =180^0

I saw a very neatly outlined connection that the atom has in its position in the universe as it was the evidence of the smallest all connecting matter tying what is matter into a small container.

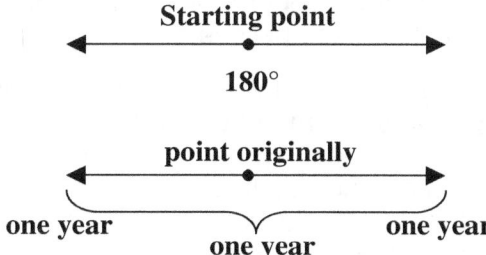

The question that nagged me for many years and on w spent almost half a year just trying to solve the puzzle is ho light travel for one year in opposing directions and after tra for one year in opposite directions be in two different poir was one year apart.

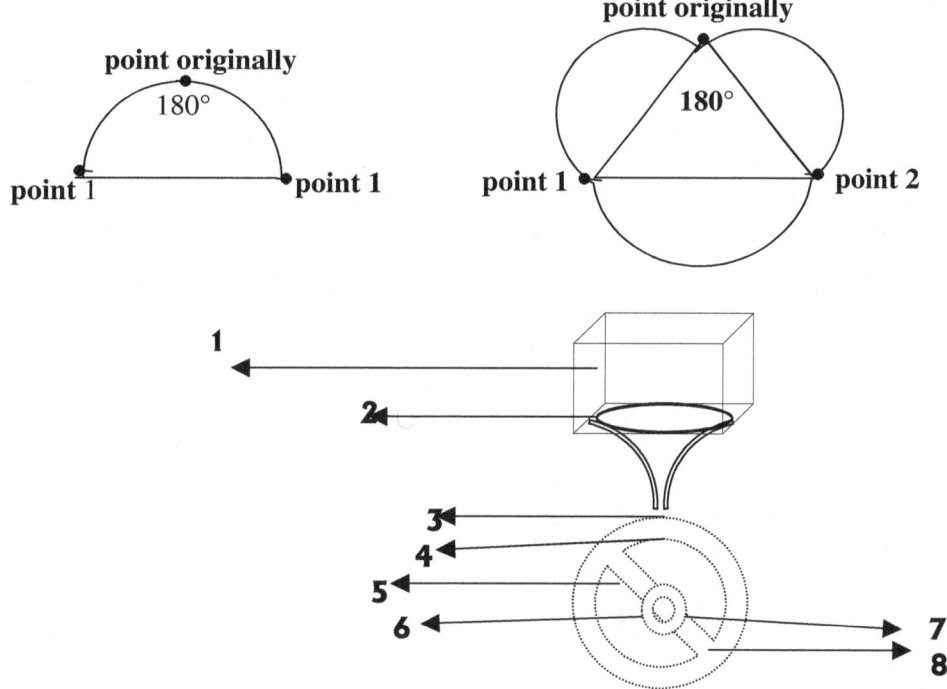

1 The square value of space-time in the fourth dimension holds a positional relevance to singularity by 10 points or places.

2 The space – time in the square of space loses the value of 10 by entering the atomic relevancy formula and become 3 sides to the cube.

3 The space – time holding space as the square loses the value of 3 by replacing 3 with Π thus going on down the line of the atomic relevancy formula and become Π sides to the circle as $Π^2Π$.

4 The atomic relevancy of space-time displacement changes once more to $Π^2$ where space goes flat in time.

5 The atomic relevancy of space-time displacement changes once more from the neutron square of $Π^2$ to the proton's double square $Π^2+Π^2$ as space unoccupied disappear and time forms the square by the square. The relevancy flowing from this figuration is so very important when archaeology presents facts and with this the archaeologists (may dare to say) became the blabbering fools they should not be as they are presenting serious science as a qualifying joke with the funnies they come up with in setting every one with thinking minds laughing.

I shall return with this argument in a later time when more facts relating to the argument are exchanged between us. For now we still have a cadaver on our hands to dispose of and quickly we must because of dire consequences that may follow if not done in urgency. Rotting corpses bring untold diseases as the great influenza epidemic of 1919 can support in evidence. Then why the danger about the corps all of a sudden if life within the corps have such minor importance as the Newtonian atheist wishes us to believe.

The molecular structure is still the same and so are the chemical composition within in the body and the mind. If it is chemicals making up man then, the man should be there because none of the chemicals went AWOL. If life is about the electricity that runs down the spine, our distinguished atheists should replenish life with some minor application of current as a means of stimulation. All the ingredients are present yet the manipulative nature of life making life so very exclusive to all other cosmic ingredients does no longer function.

Everything the fourth dimension can provide is still present within the body, yet that substance beyond the forth dimension, that ability we cannot detect but with a lot of intellectual thought, that ingredient Newtonians deny them of recognising as very special exclusive to life is not there any more. The ability manifesting as the energy or energy supplying has relinquished its role within body and mind. If it is only a matter of electrons generating pulse consistent to flowing from sector to sector in the providing of artificial current from an external source should supplement life's conducting of the work of the body. But shock the body as much as you may, the body functions disappeared with the disappearance of life from the body.

With the exit of life the vital electron distribution seized and the cadaver became just another particle containing what all substance other than life contains. The cadaver became just one more structure in the cosmos with the same ability as a rock or a mountain. The electrons conveying the massage may be replaceable but the sender and receiver and the decoding of the massage has no longer any function within the body. We may send some electrons into the body, charge the body with oxygen via a machine doing the pumping of the air to precise rhythm, as did life, we may stimulate the heart by nerve pulses artificially supplied. We may contract muscle by supply of electrons. But replacing life can never be one of our accomplishments once life has left.

All muscles including the heart and longs can be stimulated artificially and in doing it may prolong the writing of the death certificate, but having the person stand up straight once more or pronounce some wish to be for filled or just a simple effort as winking eyebrows when asked to do so is far beyond the abilities of the lifeless structure occupying space-time in the fourth dimension with the aid of all other dimensions excluding the fifth dimension.

Previously the body seemed to manipulate space-time with the utmost ease, walking wherever the mind chose to walk through unoccupied space-time, only adhering to the restraint compiled by the other dimensions and inflicted on life to startle the manipulating abilities. Now such abilities disappeared, vanished with life to some place we see as death. But life as energy has permanency that can never become denied and disappears. It cannot become washed away, wished away found to have disappeared for it is energy, a something of eternal power in being.

As energy has linking to eternity it has to come from somewhere as much as go somewhere it came from after it left. If it went latent even then it has to be stored in some place of gathering energy of life's nature. That storage facility is not part of the fourth dimension as the rest of the body is, so it will no longer be in range of the detectable, yet it must be somewhere with the linking energy holds to eternity. We also can tell that lesser forms of life will destruct the composition of the cadaver feeding and preying on it till what is left has no longer use in sustaining other forms of life as food. We learn from the esteemed and well respected Brainy Bunch the food we eat provide the energy we use. Sure that is very to the point and easy to swallow. But what uses the food in maintaining the energy. This is my problem with the SUPER-EDUCATED-WISE-OF-THE-WISE, as they will forever give information that has no substantiation but only scratches the surface and leads nowhere.

How does food give the energy and what uses the food for energy. If it is the protein filled body of the human flesh, then what makes the intake of food become meaningless after life departs. Should the food be energy then just feed the man and revive him. Food is all he needs to regain life and if such an effort fail to revive the dead then one should seek to find deeper meaning to what is obviously not obvious.

If life only connects to the fourth dimension of space in time through energy supply such as food may supply then the body is there to be nourished back to life. Also in the opening of this argument I showed how life travels time with no limits to boundaries in time. I did admit that such travelling only applied to the mind and not the body but since it is the mind that has gone absent and not the brain, then the travelling time by the mind must still be in affect as it was the mind that did the travelling and not the body. The body did the travelling of the time constraint but it is the same body that cannot even permit travelling to its grave because of the absent of the mind.

Time did not restrain the mind and if it did not restrain the mind time must have little control over the mind. Since the mind is no longer present in the body and time is still doing all the restraining on the body we may conclude with some extelegence that wherever the mind goes in storage time will hold no constraint over it. All the atoms that were in use by compounding a human body would one day again form some flesh of another being. It will ultimately not be human and it will obviously never form a body as one group of solid fleshy matter and it is logic that the compliment will become divided when forming a future body amongst millions of bodies all containing substances previously located in millions of different bodies to form millions of different life forms but it will become use full once more in the future.

With such a remark I do not say that the atoms scattered after destruction of the body of say a dog will form a dog in future. Such a presumption is madness. There will be dogs in future claiming some of the matter and there will be other life forms finding use for the ingredients that once constituted Old Neighbour but the principle is that the atoms that were in use will again find use because of the atoms eternal connection to time.

The carbon fibre is on earth placed for the use to carry and support life and as it did in the past, so it will do in the future. It is part of the eternal qualities of the atom to maintain space-time for the foreseeable future and the foreseeable future I suppose is the duration of time that the earth has to sustain life. To us humans such a concept of time on earth holds all the factors we connect to eternity because to us the earth and eternity is almost alike, but in thinking that never should one forget that in the realms of the cosmos the earth is but a flash in the pan, a wink of the eye and it is gone. But not straying that far into the future we are still measuring the chances of Old Neighbour becoming Old Neighbour once more for he was quite a likable chap and some people will miss him (I suppose).

What will the chances be of resurrecting Old Neighbour to his former self? Well left to the simplicity of the arguments held by the astonishingly brilliant Newtonian atheists we must consider it better than one hundred percent and according to their superb argumentative powers it is as good as done with the aid of a pump, an electric generator and a shovel to push food down his throat. But beware because when gauging by their record of previous successes notwithstanding the simplistic manner they go about denouncing the complexity of life, my prediction is also my advice: if you are a betting man do not bet on a positive outcome because you are about to lose money in such a bet! Your chances in winning will be as good as that of our atheists' wonderful arguments being correct.

Well now Old Neighbour is going to push daisies, or is he? Who is who and what does the daisy pushing? We know very well it is Old Neighbour that is going to push daisies because he is not with us and the "not being with us" part means gone away. Should one force he argument that his body, the compliment and assortment of DNA sells arranged to a specific order matching a pattern profile that belonged to Old Neighbour exclusively is Old Neighbour, or at least that is what our Newtonian-Bright-Boys insist on being the case. You, well any person, makes up a compliment of DNA sells and according to your sell arrangement whereby you become you and by pre-selecting sells and arranging them to a specific order where they form one totality and arrangement by assortment giving any person the prospect of life. Our distinguished atheist loses all other related arguments past this point in order to conclude what they believe to be correct. Considered in the utmost simplicity, yes, that is correct and as that alone it leaves no doubt.

Through this a rat cannot be a horse and a dog cannot be a lion

It is so very simple to understand when explained with such excruciating simplicity that even us living on the other side of the universe where the Lame-Brains belong are can accept without arguments because we are so scared of putting the least of effort into the simplest form of thinking giving the Brainy Bunch the scope of miles around to come up with the most idiotic answers they can dream up and we the Lame-Brains are too willing to accept as long as we are excluded from any form of thinking. Therefore we allow them so gracefully and with all dignity applied to both sides of the intellectual divide, to bullshit us to a stand still and make us feel great full that we were so privileged in accepting they're demising and diminishing mentality bestowed onto us. I say this from a stance where I am part of the idiots ranks and stand amongst my fellow mindless admiring those of the fortunate and privileged with they're wealth of thinking power because they achieved so many a splendid degree and are therefore the rich in thinking making me just one other poor beggar in thinking-power.

The human being as with all beings having life connected to the body structure they occupy which are the compliment of arranged sells and such an arrangement exclude my being a horse and it excludes the horses chances of being an ant.

With things that simple and sells going nowhere as they did not go anywhere in the dying of Old Neighbour why are they not functioning? What made them go on a permanent strike? Why can our Brainy atheists not once more persuade or force those sells on strike into accepting responsibility for their work responsibility because all of the world needs Old Neighbour around and the medical profession did not yet receive they're rightful chance to drain his money like a broken dam wall under the banner of keeping him alive for his family. Well at least until his medical aid runs out and his bank account has gone bust. With that simplicity being the case of life the atheist can at least replenish the life to the sells until everybody in line from the chemical manufacturers down to the cleaners washing the hospital floor had they're chance of becoming Old Neighbour's inheritors and not his wife and children. With Old Neighbour circumventing the money draining system it becomes totally unfair and what is more is why did the system spend so many billions in creating a net where they made Old Neighbour so scared of death and disease he will gladly part with all his money as long as the system gets the chance to help him cheat death (should you not believe me look at the cancer and other advertisements and think for yourself who is paying for the brain washing). Why not only tell those with cancer to do the fighting? Why charge everybody up to come out with they're six shooters a blazing in spraying lead. Who is paying for such advertisements and who receives the benefits of such advertisements all done under the banner of securing a longer life for every body.

I am a diabetic and a smoker that does no exercise of any kind but to get out of bed in the morning. I was medically ordered on so many occasions to quit my smoking, and I not sooner did that then they started feeding me anti depressing pills and anti anxiety pills and sleeping pills and stimulants to fight the sleeping-during-the-day-attacks and the.... The list goes on almost indefinite. Once I pick up my smoking habit again I suddenly do not need one of their pills to keep me "normal". While my smoke may kill me the exhaust fumes of the cars in use which pours the most deadly of gasses into the atmosphere being carbon monoxide is not maybe but definitely not only killing me but also nature in every aspect. Carbon dioxide is a natural element on earth while carbon monoxide is a chemical acid eating or more accurately said devouring even the likes of statues chiselled from granite rock as well as things manufacture in iron to rust. That aspect no one ever comes to mention BUT SMOKING is the killer destroying life by the billions! The doctors are reluctant to allow the tobacco industry to kill you because that will deny them the chance of killing you chemically and making the profit themselves either through driving their luxurious cars or stuff they prescribe and you can only purchase through chemists. I was a smoker but I stopped smoking because I couldn't afford it since I now am jobless. But I will say it is the most rewarding experience those simpleminded fools took away from me. Smoking is not addictive. Smoking is not a drug because you don't get intoxicated from smoking. I know what I am talking about because I have been through the mill and I still get hungry for a smoke. Yes I said hungry. Smoking is a system of feeding, taking in food. It is a way of taking in carbon as food through your longs just like plants take in carbon in dioxide and in monoxide through the leaves from which they grow but in the case of human consumption it is a system of feeding a person by long intake. That is why smokers that stop smoking are permanently hungry and it peckish for the rest of such a person's life. If you stop smoking then you are a quitter quitting food. Your smoking did you no harm. I am prepared to challenge any medical doctor on TV on this. I dare any doctor to organize a live debate between that doctor and me to prove smoking is bad. Smoking is a way the body feeds and not a drunk you get unrecognisably drunk on in the way you get with alcohol and other narcotic mind altering drugs and if doctors have done so little research on the use of tobacco how could they be experts on the use of tobacco. It is again diverting the money to the pharmaceutical companies because no one speaks out about tranquilliser abuse but tobacco is from Satan. To kill people on medical grounds by prescribing though controlling and minds abusing tranquillisers is much preferable especially while doctors grab some of the money and pharmaceutical companies get the rest is very preferable to people smoking and dying cheaply with a clear mind. Drug addicted patients always come back to doctors form more "medicine" while tobacco "HUNGRY" patients buy their tobacco from the corner café where the medical industry don't control the profits It is the place where the money goes that becomes this issue of keeping the notion healthy and to up the tobacco buy increasing taxes help to equalise the price difference between narcotics and tobacco. It is correct that I don't want my sons rto smoke but then it must be for the true reasons and not to feed them pills instead.

So the doctors scare the daylights out of you about death (which you will never escape in any case since you are born to die) to feed you pills (so chemically poisonous they can only sell on prescription as they are sure killers and the most dangerous available) and the system is creating another slave by making another fool so brainwashed he truly believes he will eventually cheat death! And Old Neighbour had the audacity to escape the loose of the system and die still with money in his bank account! Such a dead is outrageous and cannot be tolerated. Believe me if the medical profession got to Old Neighbour before I did, in his dying effort they would have kept him alive for another few hundred thousand reasons, reasons you keep in a bank vault and pester his wife and children with guilt so that they part with the money so willingly they will even pay anybody to advise them to part with the money. (If that is not why you pay the doctors treating a man that is ninety nine percent dead already then why are you paying him in any case).

You the reader may not see it but this is all resulting from atheism and a system promoting atheism and is an all out war world wide making every breathing person on earth a slave to milk until death does its part. Convincing people about the simplicity of life will encourage them to fork out money to be kept alive so that the slave will gladly allow more milking.

Slavery so I am told and so I do believe from the bottom of my heart is wrong. But the slaves did not have it so bad in the days of the Greek and Roman Empires. They were much better off than us the slaves of the current World Order. Slaves under the Roman law were fed clothed and accommodated on the Master's account. The law was that the owner of a slave had to feed him and provide accommodation for his slave. Then the slave had the right to ten percent of the income the owner generated from the services of such a slave while the slave had the chance (if he could) to buy is freedom Slaves in the current World Empire of the Hoggenheimers an Mammonites enjoy the pleasantness of a just system where the system does away with the need to bay slaves, the slaves join the system or die. Furthermore they make the slaves pay from their wages for food logging transport and clothes while the Hoggenheimers do not even pay them ten percent of what the Mammonites earn from their services. Under modern law, modern slaves are worst off than slaves two thousand years ago! And to top this Old Neighbour had the audacity to escape the slavery without even paying his last bid for his freedom. How criminal can a man become in such a manner of escaping what was rightfully his dues to pay. With all the simplicity about life and the promoting of escaping death why can the atheist not bring Old Neighbour back to do his last part and fill the already overflowing money caskets of the Hoggenheimers and Mammonites.

There is this wife of one certain pop star a member of a very well known group in the sixties and one of the four members in this very well known group. This wife of the famous pop star made millions on promoting the abandoning of the use of animal meat as food. She told about her and her husband having lamb chops one afternoon while some other lambs were grazing nearby. As she saw the lambs with her mouth stuffed with their friends she then and there got thinking about cruelty and the humane aspect about eating lamb in the presence of lamb nearby. She was devouring the flesh of sheep that was killed for the purpose of feeding the human population and that gave her the idea to make millions on that thought and selling humanity in the process.

For some sake of sanity let us scrutinise the situation and for once go just a shade deeper than just being prognostic in our conclusions. The lamb has carbon$_{12}$ as a mixture of forming the composition that we named protein. What will be that different from eating grain and eating flesh? Both holds life and both holds death after life. The grain is an infant that did not yet start life whereas the sheep is an adult whose life was cut short during life. Both faced death before they received the honour of completing their sole purpose on earth and that will be to feed man. She went on a campaign promoting vegetarian dishes that did not even contain fat as protein but included the biggest variety of plants imaginable.

While on the tour of promoting the eating of plants (and selling her book to millions of other fools that run on emotions they do not understand, cannot control and where such emotions totally outsmart their thinking capacity) she stopped far short of explaining why she would consider plants lesser life than what sheep are. Can the reason be that the sheep think nothing of devouring the grain and she allows the sheep to do the thinking on her behalf? Is it because grain does not run around when "chased to become grained" for food. Or could it be that the price of the book and her selling power of the content of her book allowed her to sucker some idiots (and I believe the number of idiots caught in the scam runs into millions) tinted her perceptivity so very slightly in favour of the consuming of plants that cannot make any sound or request any human emotion by running and shouting in protest trying to escape the butchers knife or in the case of plants the sickle.

In the case where we consume fruit as food the fruit we eat is food still alive in the same manner as does lions starting to eat a buffalo that is still standing on all four legs. If someone somewhere came about the promoting of eating animals while they are still alive I would surely go on the same protesting crusade as she did in her bit to fight the food supply in the form of meat. We now are faced with the same cynical questions our friend Old Neighbour left us with. Is it his corpse lying there or is it he lying there. If it is he then I have to admit that we are eating lamb. But if it is his corpse then we are not eating lamb but merely the remains of what was lamb once. I am not wasting any space on arguments about killing to eat because kill to eat we do because we have to do it. There are no other options open to us but to kill or to become killed through starvation.

The bottom line underwriting everything said about what form of food we should or should not eat is the human capability of becoming completely self- absorbed in sublimation? We think we know exactly how God created all around because we know exactly God did not create that which is all around. Therefore it is our claim to right that we may take the place of God and decide what should count where and what is food and what is not food, but for god sake keep it simple otherwise we will not understand why we may think ourselves as gods. As long as science portrait matters simple excluding the not very popular complications of thinking every thing thought through decisively we may find that being god can be a very pleasant way of living and un-complicated. If we do not complicate everything we may even think of ourselves as very clever gods without the excruciating effort of being clever gods. Just go about and visualise our brilliance in reason and tell ourselves how kind-hearted and humane we are without any deep philosophising about truth and matters of complexity.

If you are in support of the humane aspect then consider that the deed of eating fruit will be far worse when eating the unborn and defenceless or robbing the unborn defenceless seedling of nourishment so dearly and lovingly accumulated through severe hardship and unquestionable devotion in loving labour by a caring mother than a developed specimen of any specie. Remember that when eating the unborn fruit or the food meant to feed the unborn seed will be denying life the chance to be and that is very unfair! At least the meat eaters gave the sheep the feeling of being sheepish before removing the feeling permanently but in the case of fruit eaters the fruit never had a chance of feeling fruity. I should add that to my mind humanists are the worst practising sublimation because atheists deny the fact of God but humanists are in criticising of Gods way in creating the balance we know as the echo chain. Humanists are constantly trying to show all that are willing to listen to their senseless rambling how much better a job they have in mind for all life on earth than that which God established up to now through giving man reason to think with a mind and not an emotion and forgetting that the methods applied got civilization in such a tested and tried state as those methods did but still they whish to change it because they think they know so much better.

If our pop-star-wife did not have the pop star fame and all the pop press in support and with the wealth of food supply around how far would she come with the cheap mentality and the thoughtless advocating of the shameless theatrics to support her promotion of self enriching by selling books. When any nation is in total starvation as the Germans were just months after W.W 2 I wonder how many hungry men and women with children crying starved to almost death would applaud her madness as greatness. She got through because there were abundant and not because she had sensibility in her quest to make money.

She could manipulate others while the others were swamped in good times and rolling in the fat of fortune fed to burst while gloating about how their humane hearts bleed for the helpless sheep all over the world knowing very well none of them ever had to skip one meal because of want. They never had to live through one night of agony where their children were crying because the children were too hungry to fall asleep. When thinking about such conditions their gloating in self-praise is quite sickening. From me and mine to you and yours I am telling you this shocker: the total destruction of mankind may only be as far away as the swing of a telescope, and the announcement of a funny little dot that seems to grow as it is heading our way but more about this later on. She is merely one of millions making senselessly money without thought of dangers larking

This I say because nature tells the truth about man and the way mankind evolved. All predators on the hunt have eyes pointing foreword to find the maximum advantage in three-dimensional sight. By focussing in hundred present accuracy the predator can pin point the kill and act swiftly and abruptly minimising the chances of the hunted from escaping such an attack. On the other hand when looking at animals that is mainly vegetarian we find their eyes on the side of the head to secure maximum vigilance and response to such an attack. When looking at the human face we find the eyes even more in the centre of the head than in the case of an eagle, famous for his hunting skills and such a small but obvious clue demolishes the entire bleeding hearts cry for passion.

All animals dependent on meat for food sustaining have eyes pointing to the front the very place humans find their eyes to be. The road our humane idiots genes followed took them through a ancestral path with a long range of meat eaters that brought the gene carrier to what he or she is in the modern age, but being smart they make themselves the fools they are. If we humans were fruit eaters only and had no natural inclination for meat then our eyes would be next to our ears instead of being rite above our noses in the centre of our faces. Those placing meat eating in so many disputes should then also change their eyes position to the side of the head and denounce their ancestral trace of meat eating.

Man has a vision allowing 180^0 sight where as animals born to be the prey has a sight range of 360^0 and none of the humane intellectuals ever came that far in reasoning. With such direct and undeniable evidence about our eating of meat, how on earth can those shouting no in support of meat eating show their faces around as intellectual beings. This also goes to some religions denouncing the eating of meat but as long as they keep their religion to themselves without trying to convert me to such rubbish they can believe what they want and exclude me. I say this because on occasions I got into debates with such people that wanted to push their religious ideas down my thought about some Indian god living in India and you send him money with a prayer where he then fixes your problems rite across the ocean providing you do not eat meat because of his say so.

All species on earth are what their history made them. They are moreover the road they followed down to where the specie currently is than what they are at present because when circumstances change the genes with idle qualities will arouse the complexity of the specie and old habits that saved the specie from extinction in the past can come to the front and again save the members from extinction. The Sudanese can survive by eating leaves from trees until the rains arrive to bring about new harvests (although the rains never return permanently). On the other hand the impala cannot start eating lions to keep alive until new vegetation grows again. But even the harmless impala is not that harmless to grass, as grass has to grow meters every year in order to sustain the impala's nourishing needs and at the same time secure the survival of the grass as a specimen of life on earth.

Man too, if need be, can survive on grass and that puts man on top of the evolution ladder and not their misguided impulses in correcting the ways we developed. It is great to play god when God gives in abundance. It is great to play god when God brought your specie this far. But try and play god when God closed the clouds bringing rain and hunger with facing starvation. Then the mind fills only with thoughts silencing the hunger pains and the obsession comes as the hungry wish to fill the stomach with food without filling the mind with cheap sentimentality. How brave will the Super humane then be I wonder. Being humane is closing life to a very single minded approach and in this the massage of the atheists simplistic views about life ring out loud.

All this may be fair but there is another side as all things in creation stand in relevancy In my quest to find answers one question I could never find an answer for is why do the world not import the Sudanese to Britain America and Australia instead of exporting the food they donate to Sudan. Sudan has become a country that will never again support such massive numbers of people and the food will forever be needed. The growing desert claimed the country and it cannot sustain human populations. Declare Sudan uninhabitable and take the people to the countries donating the food. It will be much cheaper to feed them in the countries I have mentioned and at the same time it will please the bleeding hearts, give the Mammonites more slaves and the Mammonists more slaves to drive while not hurting the unemployed one bit for jobless they are because jobless they wish to be. Change the relevancy in the equation and take the people for once to the food and not the food on a yearly basis to the people.

Before every Anglo American starts demanding my immediate and successful castration without precondition let me add why I say what I say. By feeding the population the bleeding hearts are getting their wish but in it they are sadistic and devilish cruel. Before any aid can be requested a disaster must be in progress. Being a disaster in progress means millions are suffering. There has to be an enormous lack of food supply to wake the caller. Babies go hungry mothers weep fathers run off because they wish to find food and disappear in the process. Suffering runs deep as it runs wide and no aid can prevent that as no precautionary measures will ever be good enough. By helping once you are spreading the suffering to last longer and with more pain next time around and we all know there is a next time around because of climate changes going on. Feeling good about your self because of proving once again your good nature, your blessing heart and empathy by the giving aid helps no one because of the coming of the next time.

The simple truth is that those in power and those with influence give nothing as much as care for the helped victims. The philanthropist collect money on behalf of the Hoggenheimers from the bleeding hearts while the philanthropist encourage the bleeding hearts to donate in giving for the simple reason the philanthropist share in the spoils of the unselfish act. The Mammonites bay the food as cheap they can in names of companies they own with as little money possible from stocks the donating parties would trash in any case because of poor quality, then bay the food from their private companies with huge profits going to the private company because the selling party is also baying on behalf of the relief organisations with the money the bleeding hearts donated not because out of true sympathy, but the bleeding heart wish to kill the guilt they feel as they know they have it splendid and therefore they need to prove to all but mostly to themselves they're godlike generosity by donation.

The Hoggenheimers take their cut with excessive profits by distributing the bleeding hearts', which the philanthropist collected so unselfishly as proving it by taking their fare share of the profits going around giving the money to the Mammonites, which are baying on behalf of the bleeding hearts from their firms as they sell the stocks they previously bought for next to nothing with excessive profits. At this point the Hoggenheimers bring in the Mammonists to do some slave driving as the spoils has to be sent across the world. In this heart braking act of generosity some more unbelievable profits go the way of the Hoggenheimers and the Mammonites because the firms involved just so happens to belong to a shared venture between the Hoggenheimers and the Mammonites and by some more overwhelming generosity they share crumbs with the Mammonists doing the slave driving.

Now you tell me who is unhappy while all this good heartedness goes around and is there any blame to be where the rich becomes richer as that is no one's fault. If the bleeding hearts were serious about their conviction in generosity they would not bay some guilt relief. If the other parties were serious about their convictions they too would try to find a permanent solution but then there will be less profits to gain. The bleeding hearts are quite satisfied that big planes are used to transport and distribute the food but they know very well that that is the most expensive means of transport and someone somewhere is changing very unselfishly a dime spent to a dollar wasted. In this way the relevancy is getting the rich richer, by giving the guilty guilt relieve and helping the luckless to another round of heart ship in hunger.

Change the relevancy around if the act is in pure kindness and brother love. Take the luckless out of the equation of desperation in cycles by removing them from the problem. In that there are some more relevancies involved. If the bleeding heart were serious they would never mind bringing the luckless to share in their abundance. The other part of the option is to let nature take its toll rectify in natures way and be done with it but then the profit issue stands to lose millions of reasons why neither option is an option.

The relevancy will lead to a cheaper solution although more expensive the first time around. Everything is about relevancies. On the one side of the relevancy is the earth became unsustainable to carry a human burden in that part of the world and on the other side of the relevancy is, the western countries have food to donate in tons through baying and selling agents, (and I shall gladly eat my farm if the politicians were not sharing in the bounty of tax money donated in generosity).

The one side of the relevancy is the Sudanese will never be self supporting because on the other side of the relevancy is in the long run a desert means drought and water will never again be abundant. The only solution to the equation in solving the problem is by changing all aspect around in the relevancy and through that finds a permanent solution to an unsolvable problem that will forever remain unresolved until the relevancy changes to finding an answer instead of avoiding a solution.

If the cosmos can tell us one thing it is that changing the relevancy brings about solutions. By creating the Big Bang it solved a problem of overcrowding as we have in Sudan and by creating space as we should in Sudan the cosmos separated matter from space as it is still doing with the Hubble constant proving that space is on the increase. But if space is on the increase and all is about relevancies something else must be in decline on the other side of the relevancy to find equilibrium between the problem and the solution.

The cosmos brought in space on the one side and matter on the other side and between matter and the factor of space growing must be some sort of problem solving. If we wish to find the answers to the cosmic mysteries it should be the most obvious starting point because there is one side of the relevancy known to man and then looking on the opposite side of the relevancy must be the solution. Where one thing is growing something else must then be in declining and in that comes the answer of the relevancy that I share with the introduction of my theory on matter holding space in time.

Most prominent in all relevancies there are must be the atom, the one little container giving matter character and different uses in the universe and by adding or removing one small part it changes in character, as Doctor Jackal and Mister Hyde never could. Every one knows what is in the container but what is the container in? If someone ever gave that thought the light of day I have missed it. However we also can learn much in thinking by placing a reference we see with life back to nature. If we cab see light by using a photon we pick from space, then we could only do that if we copy what happens in nature in using the atom.

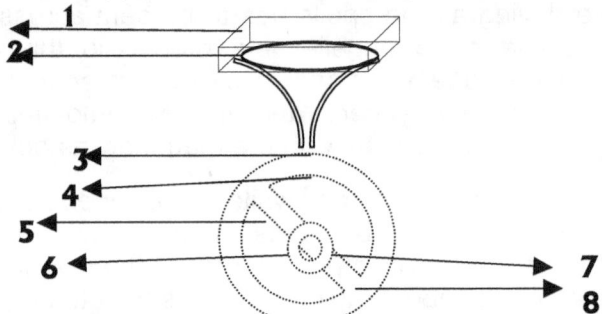

Referring to relevancies means one may exclude only nothing from attachment and as such that put whatever there are to consider in a relevancy to whatever there are to consider. The universe is one giant spinning machine holding everything in tune and aligned with all ells. Everything become relevant as all in the cosmos divert from singularity by the line of singularly applying sides to singularity forming the triangle in singularity bordering the half circle to form a position where Π will become r^2 and lead on as a value of C. In singularity the value of space held by lines diverting from singularity forms the space value of $\Pi\Pi\Pi$ which is so close to eternity the time value applying exceed the dome compliment of Π^2 matched only by the half of the square being the triangle at Π^3

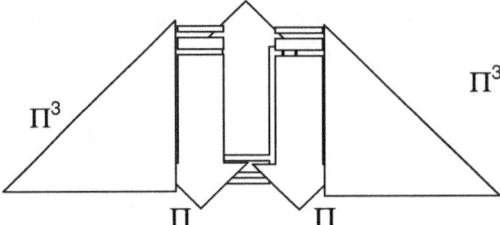

Diverting from singularity as extending Π time claims space from singularity comes about the double proton in space $\Pi^2 + \Pi^2$

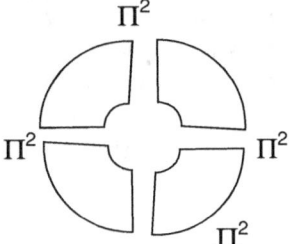

From the square of Π^2 the dome as a half circle by four places the line implicating the line by four to place the triangle relation by double half a square.

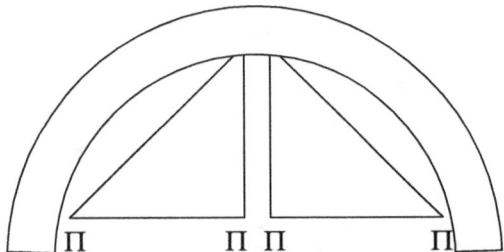

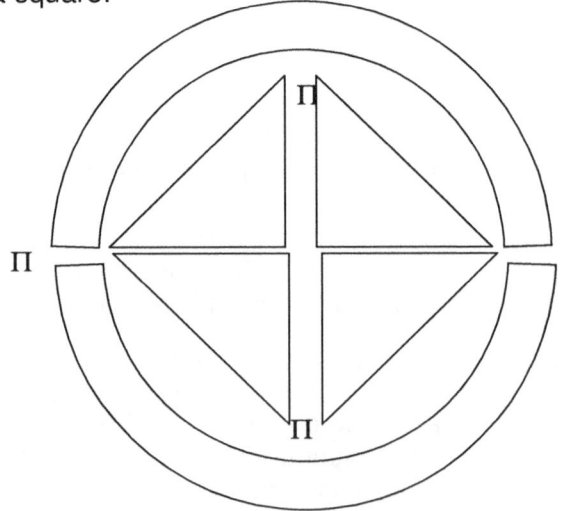

The total value of the four in the circle bringing about the space of the triangle reflected onto the time the circle holds committing the straight line to a value becomes the fourth dimension of space-time.

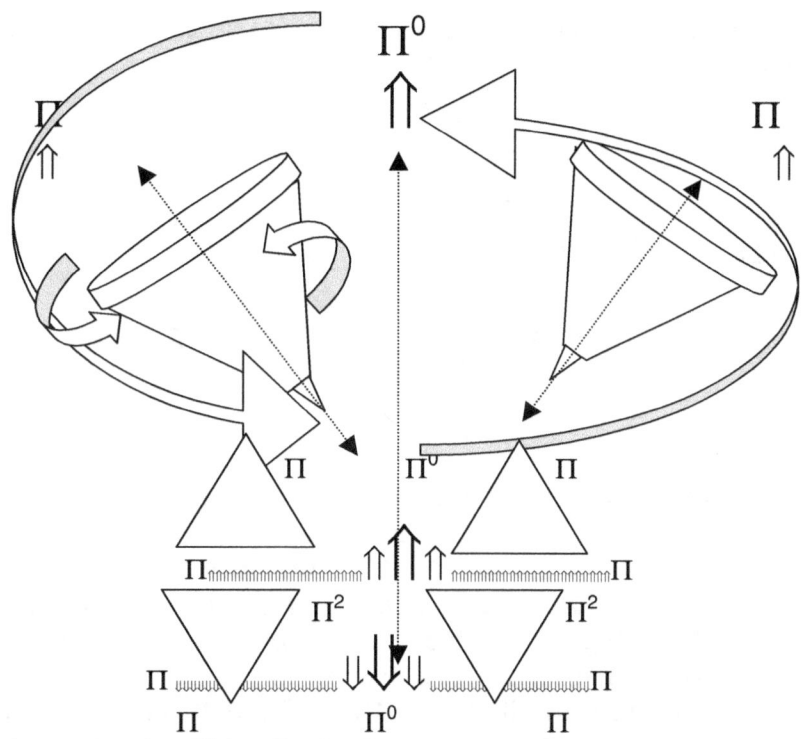

In this it is clear why the Titius Bode ([10 + 10 + 1 + .991] / 7) and the Lagrangian 5 \\ 7 systems part their ways when applying the different processes they hold. With all the differentiating, the observer must also consider the dual massage that light uses in travelling through the vastness of universal space. The thought of nothing is just what it is, a thought of nothing and although it is in the human mind common nature to present nothing as a value in the recalling of something, nothing is a presentation of the figment in the human mind. There can be no number such as nothing and that was (possibly) Newton's biggest error. Nothing represent non-existing and that is just what nothing is, it is non-existing.

In order to prove my point I wish to ask the reader to define the shortest line there can theoretically be. If he should answer anything but that the shortest line will be at a point where the beginning and is the very same spot he will be wrong. The shortest line that can ever be anywhere must have a start and finish holding the exact same spot. The line will be humanly impossible to create but we humans are capable of very little.

Stars can and stars do overheat, sometimes and the polar regions where the Titius Bode matter to matter applies holding the square of space (10) in a double relation to the square of time (7 + 7)

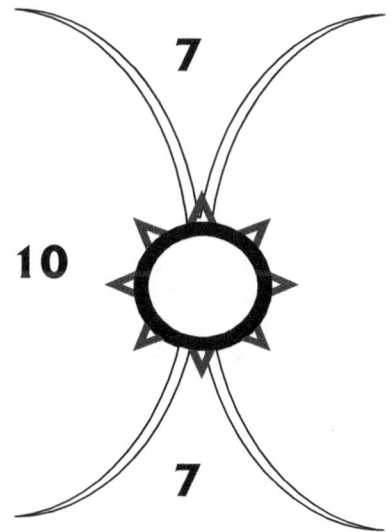

Other times they overheat in the Titius Bode principle holding the square of space (10) in relation to matter (7)

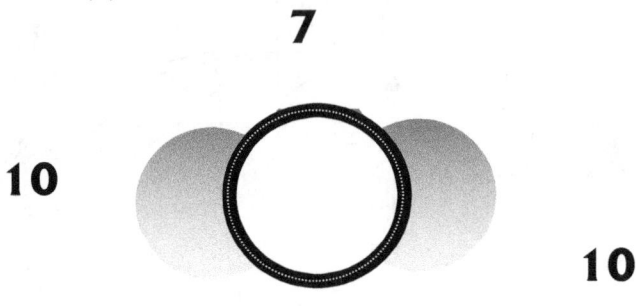

This comes about through the overheating of singularity (7 + 7)/10 (top) or layer overheating
10 / 7 (bottom)
It is no longer an issue that stars overheat but the issue shifted to the question of why stars overheat. Applying the Titius Bode laws (shown above), the Roche and the Lagrangian principles correctly, I can prove that :

2) There is no gravity and therefore GRAVITY DOES NOT EXIST.

3) With no gravity it stands to reason that I also maintain that both NEWTON AND EINSTEIN ARE wrong about their views on cosmology.

4) With no gravity NEWTONIAN VIEWS ON THE WAY as how CREATION CAME ABOUT IS HALF CORRECT BUT HALF IS INCORRECT.

5) And shocking as it at first may seem but true enough is the fact that THE BIBLE IS CORRECT ABOUT CREATION.

Could any one ever make claims as I do and at the same time claiming sanity as being one of my virtues. Well I shall try to explain how humans see the Universe in total contradiction to how animals see the universe and to top the lot how atheist and animals see the universe alike.

Animals use eyes but we humans have more too see with when using what we have by using our minds to see with. We should see the universe with the light of understanding shining in our minds. When one look at the night sky one see darkness with little specks on light. Why would anybody see darkness because darkness has no light. Yet we see the darkness. The darkness should be invisible if we are seeing light because the one contradict the other. If the night sky was black then black is what we should see but then again black is the absence of colour and colour is the visibility of light.

We see the darkness because it has light it is withholding from us and while withholding the light from us we see the withholding as darkness and the darkness we do not see because we are not suppose to see darkness. That makes the darkness we see not darkness to be but it is in fact light we can see as darkness. On the other hand there is the brightly lit dots we can see because the light shining as dots are darkness as the darkness are stars giving us the light they are not withholding. As they are not withholding the light but pouring it into the vast container of light the stars then become the darkness we cannot see because they give us they're light and by giving us they're light they then have no light to have. That means by giving us they're light they withhold they're darkness from us and that makes the stars filled with darkness. That means what we see as light is light and what we see as darkness is also light instead of what we ought to see as darkness because that we cannot not see is the darkness.

Therefore when God gave the command "Let there be light" it was the command "Let the universe begin" because the universe we see is light we do not see and the stars we see is the darkness we do not see. When one is looking at the darkness as an animal you will be seeing the darkness because the mind you use is that of the animal. Then, yes you may be an atheist because all animals are atheists. I have never herd of one bleeding heart or philanthropist of whatever kind convert animals to any religion there is available. If you are an atheist and you see the night sky as darkness that would mean that is what you see as an animal; a darkness that if you had the sense of a man you ought to know that it is impossible to see darkness therefore it must be light. On the other side of the relevancy you also ought to know that by seeing the light the star is giving away the light and when giving away the light it has to hold all the darkness it is claiming for own use because only by claiming back the darkness can it give the universe the light as it does otherwise it would give its darkness by withholding the light. The darkness is singularity uncommitted to specifics, spinning at the speed of light never pointing in one direction long enough to shine as light but shining long enough not to be darkness. But being light we can see it but because it is in random spin, spinning at the same speed we use electrons to convey massages when translating information we cannot see it. The light then becomes darkness because it is extracting all the light through the one singularity line uncommitted not energising it because of the absence of a replacing source converting new light. In contrast to that is the light we see because of reasons forthcoming from not being able to see darkness we can see it as energised uncommitted singularity with the aid of a sustaining in singularity from a committed form replenishing the uncommitted singularity to maintain direction.

I do not see how one can be an atheist and put claim to being a human while observing what there is out there in the way animals observe by only and purely relying on the eyes without incorporating the mind. I cannot see how any human claiming to be human cannot see past the barriers restraining the animals from being human. With minds it is so clear what the Word of God says, but to be human and not see what humans should see is a dangerous reflection on the mind you use. God did not say "Let light be visible" or "See the light", He said "Let there be light" and that is what there is. If humans then see darkness where they know that one cannot see darkness the darkness are within their minds and therefore they are atheists not withstanding that they may or may not claim faith as part of their thinking. If you cannot read the Bible through human vision and as a consequence not understand the Bible don't blame the Bible for your inabilities but blame yourself and your inabilities. It is not the Bible you cannot read it is you that cannot read the Bible. Place the relevancy where it and as it belongs.

We are human therefore we have light in our minds and ought to make use of that! This very afternoon as was writing this part I took a break and lo and behold, one of my sons came to me with a problem of a religious nature. I shall not go into detail about his problem but I asked him to define religion and what life is. To strike some sense between his problem and the size he sees it in I asked him to tell me in his view about the contents of the Bible according to the Bible and the dominie (Afrikaans for preacher man) how would they define life because some parts in discussion about his problem was the discussion involving tackling the issue and thereby the issue turned to how far can you go in solving matters and leave the rest to preying and doing prayer. I am of the opinion and will die by that opinion that prayer only serves a purpose in thanks when you yourself completed the task without preying for some force to help you complete the task at hand. Life is the manipulation of your surrounding and that means you do things yourself if you want things done and you do not prey for things to be done on your behalf by God. That is the definition of life. It is the manipulation of space-time and involves neither magic nor divine prayer but you go about changing your surrounding to match your needs. What all preaching never advocate is that we are in the seventh day of creation where that specifically states that God went to rest and from that I draw the conclusion man can and man must do everything by himself because God clearly says He has gone to rest. We do things on merit by ourselves or not at all. That is the energy we think of as life. The fact that we have the ability to self-sustaining and not being fixed to the universal position space-time landed us in gives us life. With life in hand you manipulate what ever you can as you replace positions to suit the required changing of objects where changes are needed. Then your acquired needs changed them to be to your taste and there is no other way out. Life is about changing your surrounding for the better of yourself or others and to improve all around you. The ability to manipulate space-time is the energy I have and Old Neighbour lost. Still it is energy. It is neither food nor electricity but it is a more advanced form of energy than the energy mentioned. Another part of life is tacking the responsibility for change your manipulation may bring about and the effect such change may have for other beings sharing space-time with you. Never confuse the needs of others with needs of your own and project such needs about yourself as beneficial to others without consulting others. This is very typical human behaviour.

With the Newtonian confusion raging man has mixed matters bringing about a highly unsatisfactory climate where we try to pin cosmic value and pre-conditions on life and place very stringent condition suitable only for life onto the cosmos. That leaves science in disarray and confusion. Heat sustaining life as pre-condition Xepted science projects to stars and where stars fade we allow them to die as if blessed with life's changing and renewing.

Stars certainly do not have emotions and when they erupt it is not in anger. The chemicals stars need to maintain singularity is very poisonous to man and the matter making life sustainable will have no chance of surviving even as a flash in the star. We think of a star being hot in the manner we translate life's pre-condition to what is hot. It is to the letter the same way that we take outer space as being unsustainably cold where it is quite the opposite applying.

While looking at the earth we think of the cosmos. We reflect what we conceive as conditions to match life being normal to the cosmos. Planets have to be plenty full because even we have one in hand and eight others in the back yard as spare should we make this one we have untenable to life. And should we run out of planets to ruin there then should be others nearby carrying life on one in nine, as is the case with us. We try to find life everywhere because life has such abundance on earth in everything we see. We even reflect our vision of time to mach time in the cosmos giving the start of creation an earth bound time range never thinking that the universe is growing and not dying. In the same manner we think of the universe as a living organism while the universe constitute every aspect we relate to death. In fact, the universe is the ultimate death. In the universe everything will only be once and never again whereas with life there was as much as there will be and even more will come than what was. That is the last thing one will find in the cosmos. If time ran out for whatever time will not replace or bring back what ever. Even the way we portray the earth's surface we wish to reflect to space using the same methods we use on earth. One mile will be one mile wherever you wish to take the mile. After all one mile is one thousand seven hundred and sixty yards (if my memory serves me correctly because this is still part of my culture when I was at school and South Africa used the British yard stick). Not once comes the thought that man cannot step one yard in space. Still one yard will be one yard wherever the yard may follow man. Man has acquired the inability to divorce life and the cosmos for some reason we can presume as cultural. Unfortunately we go in accordance to what we see and that is more cultural than culture it self. We see a shining light and presume it is a star in the same manner as we see a large dark antelope with horns on it head exactly in the same way as that of a buffalo and presume that what we see is a buffalo. In the case of the buffalo the past thought us such observations are correct and hence we grew accustomed to the culture of believing our eyes.

Never do science take charge of thought and divide flesh from energy in the manner I have done during this the writing of this article. Outside the view we have we can locate a something that is there but needs some vision in extelegence to locate. It is a small part of life that has an attachment to the physical but an overwhelming comes attach to something indescribable to define.

In other books of mine I try my best to prove that our view of trailing outside the sphere of the sun is a myth and even travelling to another planet is not the same "as going abroad". There are so many dimensional barriers attached to what we can see without our locating or even knowing of such existing barriers because they remain unobservable barriers. There are so much more than what ever may meet the eye.

In part 7 of the Theses I touch on the subject about the age of the earth and how short sighted (once again) the Newtonian view are on this matter. The earth is in truth not 4.5×10^9 years old but the core was part of the cosmos during the birth, the very first moment of the cosmic birth. Many processes came to change and shape the earth to what we enjoy today, but the inner-core-value came from the first parting of the singularity Alfa.

How life started as such I do not wish to speculate on, but logic tells that what ever was at day one of singularity Alfa, nothing since was added or removed and that puts the carbon carrying life at the very start as well. It would be reasonable to suspect that all cosmic structures holds the carbon but not all structures can present a satisfying environment to sustain and protect the singularity of the carbon in order to bring it to a point of holding divinity secured.

One opinion that I strongly hold is that Chandrasekhar is as misinformed about his carbon-a-plenty theory as he was about his crushing stars in weight. Carbon cannot come from the cosmos and go through the Π limit unscathed to infest the earth. That is as Newtonian as all other bullshit can be. Life in carbon was a part of the earth as it was part of the sun, but it had its being burnt to blisters and could on that account not develops on the sun.

What ever the earth went through was also a survival test for species on earth. What ever the sun threw at the earth the form of life that was dominant then, had to make do or die. The fact that life made do is testimony to life's survival skills. Life will last, no matter what man may throw at it. It is man that places man in jeopardy. Man is the prize of life's achievement that I do believe. Man is the accomplishment all other species carried the burden of. Life is built into man and all qualities of life manifested in man.

That makes man the youngest and the least protected. That makes man the weakest link in surviving. I have my sincere doubt about modern civilised man's ability to survive even the onslaught of a brake down in civilisation. One harsh winter and not one in a thousand would be able to see the next summer rains bring relief. Picture a big city without electricity for one month and think who would survive even such a limited test. One hundred years ago such a remark would have made me as silly looking as the claim I make about gravity. But man has gone down the tube, at the end of the ladder although to man's thinking he is at the top of where he ever was before.

We are launching a chemical war at all pests we do not seek. We kill and destroy them without thought. Bacteria, fungi and, viruses have been at tests far grater than man can produce and survived to tell the story to the next generation. It is written in their life code for the next generation to read and fight. When a species are at it greatest danger of not surviving an onslaught on its very existence a factor much dormant in normal conditions kick in. That factor rewrites the coded massage and the following generation find armoured protection. Man is weakening with all the chemical aid we see as medicine protecting us while we put the most dangerous forms of life on a survival course we cannot afford the luxury of. The day will come when there is no stopping these killing-surviving machines and we, man will stand defenceless while they go about killing and maiming on sight. Every little headache is a call for aspirin. Every cough is a call for anti biotic. One day we will find the disease and ourselves defenceless well and truly developed. Man will die and the count will become more than man can destroy human bodies. That will leave corpses for more viruses to grow and plan more attacks. Payday has to arrive we must see that coming and not be as arrogant in our self believe.

The fashion of the century is to place all, as equal and life holding space in a dog is equal to life holding space in an ant. That can never be for the single reason that all life in the body of a human cannot be equal. Any person can go without a limb notwithstanding the sacrifice they endure in whatever function. Losing an arm or a leg does not risk life at all on the condition that it is removed before it may infect disease to other organs in the body. Losing a liver is serious but machines may provide such an organ function replacement and life goes on, fairly difficult but without eminent danger of death to the rest of the body. The same argument can be said about the heart longs kidneys and such. The function organs play in maintaining the body is crucial but not vital. Losing such an organ does not mean death by necessity and can become even to some of minor significance. When losing the head or part of the brain things turn to a lot more serious nature.

I have witnessed friends of mine that were motorbike maniacs like I am, falling off their bikes and receiving head injury. After the recovery those persons changed in a manner where they became alien to themselves. They became another person no one new before and none can recognise. Such an injury is very serious and lethal, more to the persons that love him than lethal to himself. The persons that love him has lost a love one and gained a stranger they do not care for. Even in one body all life does not stand equal let alone from specie to specie. Losing my arm is not the same as losing my life because I can still live (more unpleasant but that is not the argument) with such a loss. The conclusion of logic is that the arm is not the "me" I lose, as did Old Neighbour when he went missing leaving his remains behind. Some of my body is life in issue for use to be discarded when no longer required for service but other part is much more closely connected to me as life.

This brought about the atheists campaign that life comes as part of a wholesale package wrapped in a carbon container and all philosophy centred around this argument went missing when some connection was proved between electricity and motorized motion of body muscle and fibre. This was the dawn of electricity and the wish-wash that went around with miraculous curing by only sending impulses of electric devises that could cure all and almost bring death back to life. Some devises remained proving through time their worth but in general it was a lot of quack and most disappeared where they came from.

Then came the theory that life was only electricity flowing from the brain to where ever body motion required the flow and all other philosophy went silent. It is not hard to imagine why because physics place electricity as a force with the same presumption (though they will die before admitting it) that a force has a control in similar fashion to a ghost or some unknown free spirit running around to every one's amazement. That mentality sticks like glue and much of that influenced scientific arguments to be in apathy to the philosophical and since 1945 when the physics got hold of the German nuclear bomb and let it loose on Japan it is mathematics ruling logic to the point of madness. No one since then had any inclination to touch this aspect again since all were satisfied that everything was flawless. Flawless indeed but at the heart of mathematics and in the very start of physics lured a flaw that became more apparent every year and the flaw eluded every one to date. It even diminishes all sensible argumentative possibilities to a stand still.

Losing a limb might not kill and it might not change any personally but it is loss to life. If some one acts promptly and in time doctors commonly have the ability to connect the lost limb and with some minor complication the limb may even restore to normal application. Would such prompt action work in the case of Old Neighbour being officially dead for say twenty minutes. The answer may be yes and more likely no because it depends on the brain damage that occurred in time laps where the brain fibre were starved of blood and more important oxygen. As was the case with some of my biker friends brain damage can and more likely will result in a mild to drastic personality change and in some cases dangerous insight attacks may occur.

Changes of such a nature are very serious and symptomatic of injury to the brain. In the brain damaged victim likes and dislikes behaviour pattern and mood swings will change the personality of the individual. The changes may result from a blocking of the flow of blood and it may result from a nerve area that lost function culpabilities but life still remains present. From physics point of view I am of the opinion that it is a natural phenomenon gone very bad and such changes in personality takes place with or without injury. The Romans believed that when a person breaks a mirror he is doomed for seven years because the broken mirror damaged his sole. This we modern people know is just another folk law tale but with some angle of truth. Of course the mirror part is the untruth but there is quite some truth behind the personality changes with an interval fluctuation of seven years. I would not go as far as putting a stop watch to the date in seven years but in a more or less manner we all show some changes in personality and a man of fifty will not find the company of a few teenagers to be friendship bonding and neither will the teenagers like a fifty year cold going gallivanting with girls very pleasant. Of course once again there are many exceptions to the rule and as with all else in the cosmos there are relevancies changing circumstances that may occur. What is without doubt is that the link between life playing a part and the fibre connection playing a part and it will be as silly to claim the carbon has no influence on the life energy as it would be to deny that there is another energy present above and beyond the fibre. With this I wish to introduce my Theory on the Seven Dimensions and I put it to you as I originally started with without changing some of it to fit my present day views.

1.4 THE SEVEN HEAVENS

Although from the name one may have the idea the article is exclusively attached to the spiritual as much as it is about religion and has nothing to do with physics. When a friend of mine saw my article in one of my scribbling pads (this was years ago before any idea of writing a book ever entered my head) he was astonished by my claim that it was pure and unadulterated physics. This was my advance from nowhere into physics. Justifiably you may say as my friend did so many years ago that the seven heavens have no bearing on physics but by saying that the biggest mistake comes into the open. I admit whole-heartedly I did not realise the importance it had back when I wrote down the loose ideas but in retrospect that was my initiating although not my first ideas.

Every aspect of every aspect connects in some way leaving only nothing unconnected. It should be somewhat obvious by now that I see "nothing" having no claim in any form of nothing as part of mathematics or physics and to my view that is the main difference between arithmetic and mathematics. In arithmetic there are an allowance for a number or a marker such as zero or nil whereas in mathematics no such number can be found because no such pointer can claim any position from the origin.

Even when I wrote the thoughts down that many years ago I did not yet dispute zero as a number, but I have to admit I had some difficulty with the value of nothing. For instance what was more nothing and what was less nothing when there was two of nothing facing each other. In all of mathematics there has to be growth as much as there has to be decline from wherever any marker may be. In the article I show that the line the half circle and the triangle have on common factor in as much as all being 180^0

A straight line cannot start at zero and still be a straight line because zero extending to wherever brings about a full zero. A straight line starts at the point where the pen point meets paper. That point may be any distance from infinity to a measurable dot, but it cannot be zero.

180^0 X 2 = 360^0

Any straight line is also half a square be cause the line forming the square cannot start at zero for the reasons I just mentioned. That is singularity pointing an eternal direction from a point of infinity and that is the basis of the cosmos as much as that is the basis of mathematics. To escape from nothing one has to become something and by doing that one could not have been in nothing in the first place. If one holds a point in nothing one cannot become something because of the nothing value.

To back this argument that no line can ever start at zero is to ask the simple question: what will the length of the shortest possible line be. It must be a line where the starting point is so close to the ending point the distance parting the two is incalculable yet there is the line therefore the end and the start is apart still sharing the same spot.

The difference in the circle and the square is the direction the indicator follows and a square cannot spin, as a circle cannot be motionless The factor of Π indicate eternal motion and NOT zero motion. There is a massive difference in that concept. If no line can have a zero point to start with where will the circle get the zero to indicate motion! This principle is the most basic mathematic rule The method applied when calculating a wave is by finding an average in the triangle continuing from the straight line to the pitch of the wave and then the decline will form a duplicate presenting the other side.

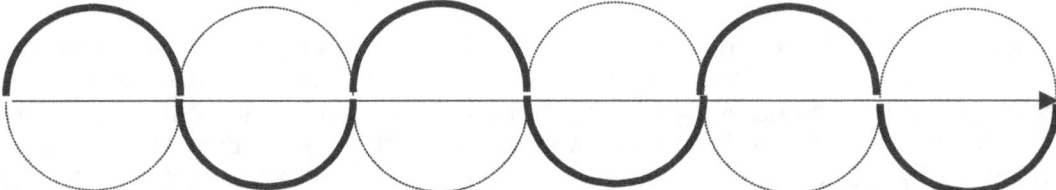

When the end of the rotation arrives the end rotation also announce the beginning of another rotation and not nullifying of the previous rotation because the rotation will have a line showing the effort it made and as it forms a wave, the wave will be there forever. The pitch may decline to a straight line, but the line remains. The wave confirms rotating directions followed by the circle as it spins. By stating that a wheel has a relevancy of zero by completion of a rotation such a claim denies the wave its rite of existing. The wave going flat, as it becomes a straight line also has an indication to singularity.

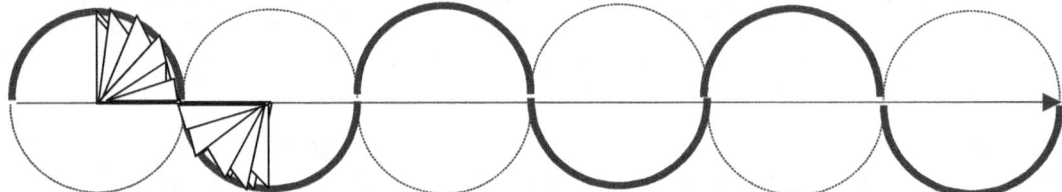

Being a circle means the thing must be round and spinning. In that case, let us take an example well known to all, the spinning top. The top spins on the thinnest of points, and still maintains a balance. By being a calculating value to match the work done in the rotating half circle the triangle depicts the flow of the straight line.

The straight-line holds a duplicate value of $180°$ to the half circle as well as the triangle all being part of singularity as much as being positions from singularity. That alone has to confirm the connection existing in the dimensional aspect.

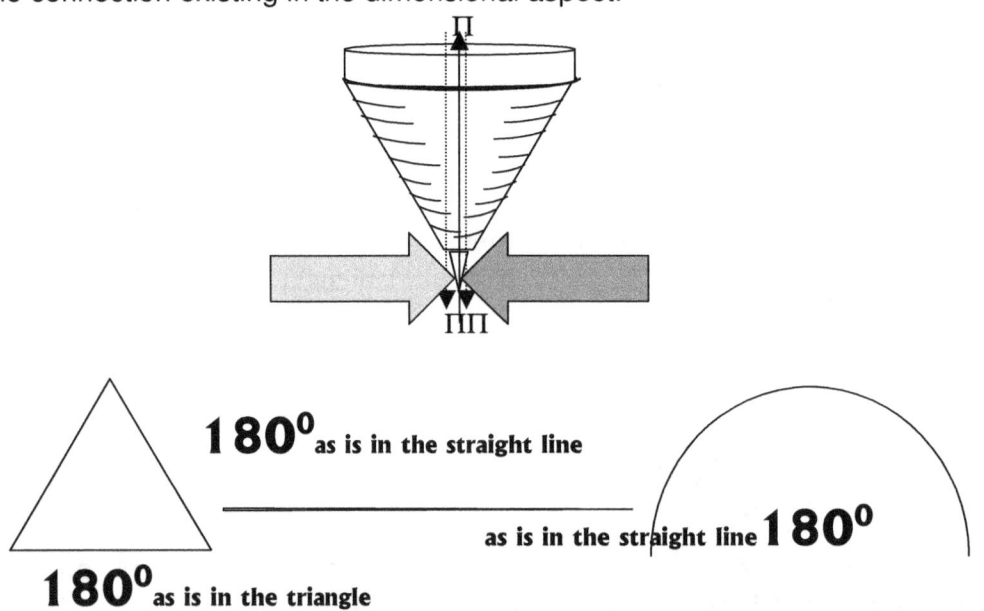

$180°$ as is in the straight line

$180°$ as is in the triangle

as is in the straight line $180°$

The dynamics behind the two principles is much, much more complicated than what the illustrations as shown above would suggest. However by using such basic of illustrations the simplicity might be tending somewhat to come across as misleadingly simple, but taken down to the core of factors behind the principles that forms the most basic of the principles, the illustrations prove rather effective in explaining the crude idea. However, please do not be fooled by such simplicity, in the very detail analysis it is as complex as can come.

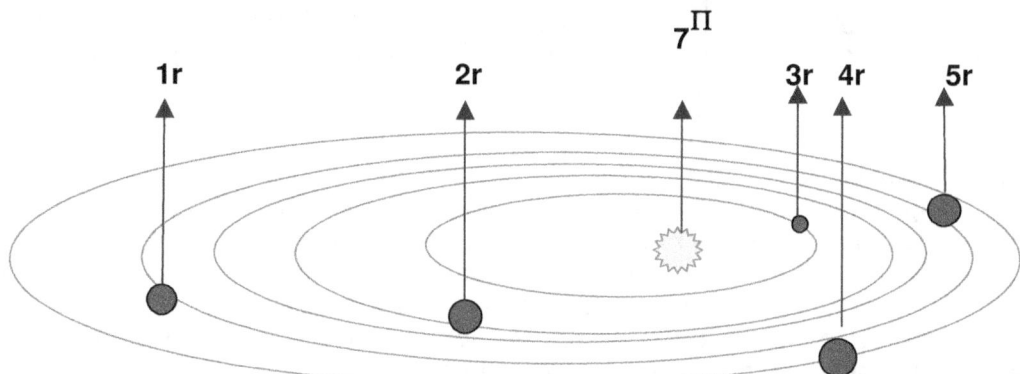

From the star holding a dominant point or most valued point in singularity it affirm all five other structure each holding singularity individually.

The universe link in so many ways we will not begin to realise the manner within the next thousand years. Electricity is one part of the link, but there are other links we may never come to know about because there is always another part of the cosmos above and below our perception and abilities that will elude us.

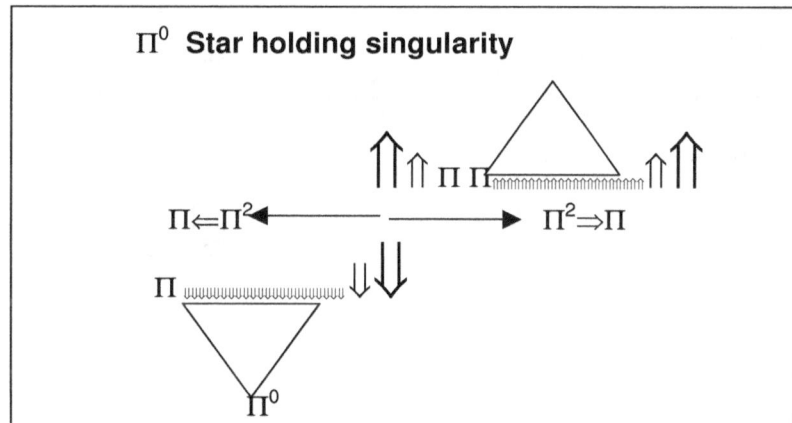

The network of individual singularity not only provide spinning through governing singularity in the sphere but also provide spinning in the geodesic through out the cosmos linking all matter to matter in a network no one will ever come to understand in full. In the sphere the foursquare triangle holds space in time maintaining singularity of different assortments. In view of the matter-to-matter Roche factor where the factor consists forming relation between particles occupying densified space-time of where ($\Pi / 2$ X $\Pi / 2$) relating to the foursquare triangle the value of gravity Π^2 comes in position as $\Pi^2 / 4$ X $4 = \Pi^2$.

A STRAIGHT LINE , TRIANGLE AND HALF A CIRCLE WILL ALWAYS HAVE EQUALITY IN DIMENSIONAL CAPACITY PROVIDING EQUILBRIUM BEING 180^0 BECAUSE EACH ONE SHARES A COMMON DINOMINATOR IN SINGULARITY.

As the straight line averts a zero it holds another straight line in place to set about such an averting where the two lines will always carry a relevancy in elation to progress (the triangle) and a common denominator in the start from singularity.

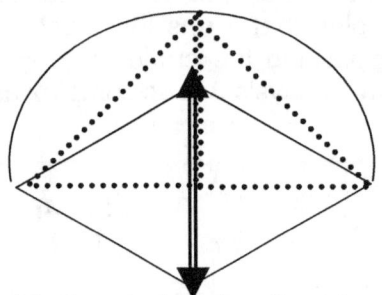

> With the normal extending of singularity it will always form the triangle in a half circle whereby Π relates to the cube by 5 points to either side of the line singularity forms. Thus there are 10 standing related to seven and visa versa.

As singularity holds the straight line the triangle and the half circle as a base to form giving all and everything next to connected to and adjoining any form being of a straight line half a circle or a triangle forms space time. From singularity in the straight line (180^0) the half circle(180^0) and the and the (180^0) triangle matter form space in holding, claiming space by controlling space to influence space, but as maintaining singularity insist on space in spinning to the time singularity dictates time sets from such spin motion and by diverting from singularity time forms the law of Pythagoras in the square of space –time.

The normal flow will allow singularity extending to 10Π but when singularity blocks another sphere in singularity the two will form a joint value and by joining the larger will dominate the space as well as the time of the lesser taking control of the surface and the atmosphere. Through this the Roche lobe comes about with all its other dynamics I describe farther on in the theses.

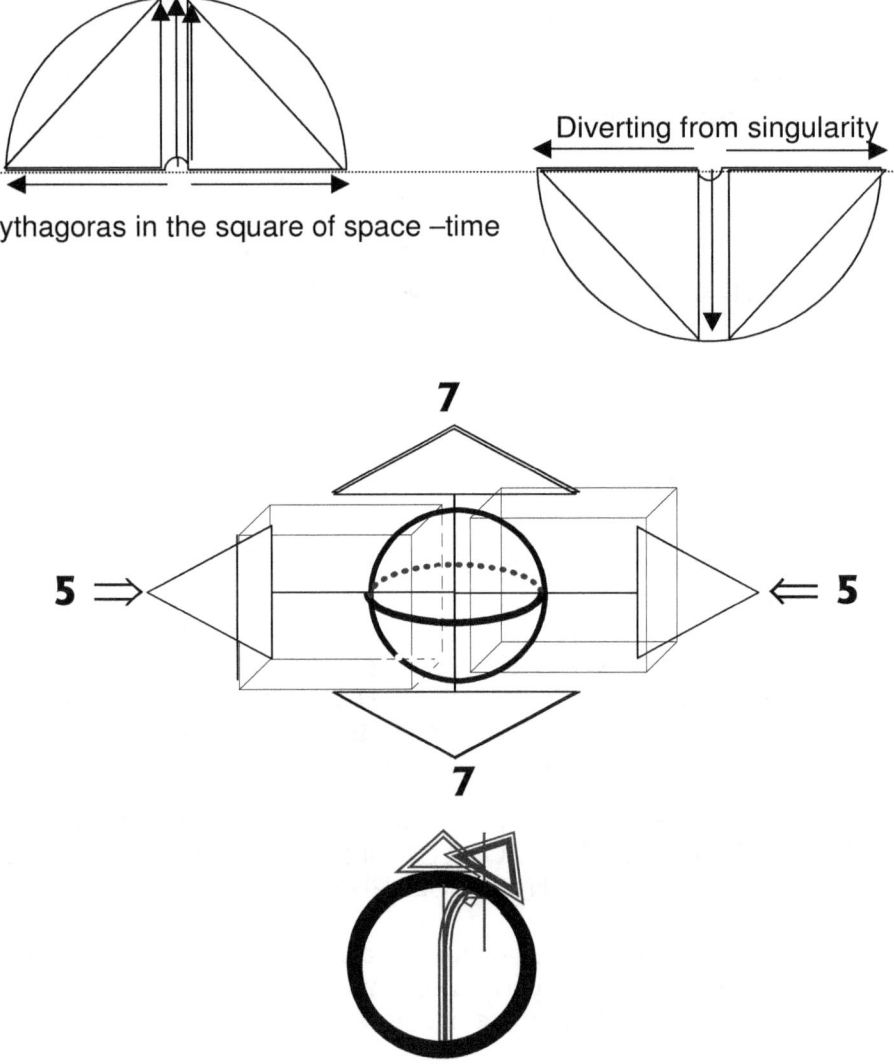

The result of the seven markers that matter diverts from singularity is present in the 7^0 inclinations the earth holds as a sphere as all spheres have. Matter is always moving seven points way from singularity as it progress in space through time.

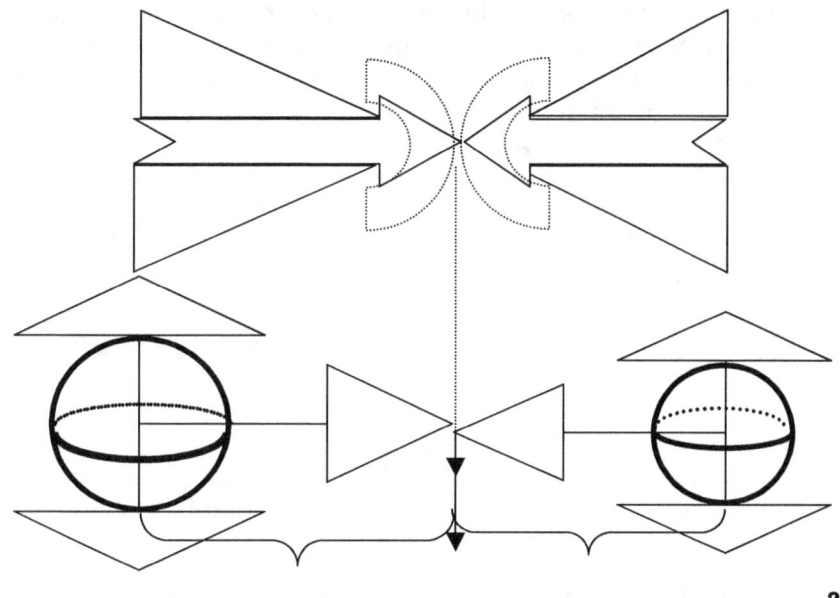

$$\Pi/2 \qquad \Pi/2 = (\Pi/2)^2$$

SINGULARITY MEETS AND COMPLIMENTS EACH OTHER.

The diameter of the cosmic structure holds the value of r and singularity holds the dimensional value of Π meaning that the radius or diameter (r) extends to become the diameter multiplying the value of singularity. But since r already consists of the square of space holding a definite positional relation with the value of singularity being Π the diameter comes into effect.

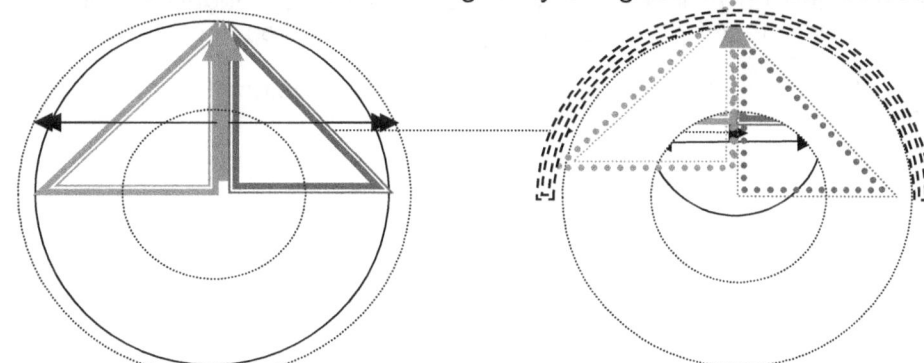

At this point the equality of the straight-line dimension to the triangle and the half circle holds prominence as a straight line, a half circle and a triangle is dimensionally equal. The common denominator will bolster all factors to an equivalent ratio,

When singularity by the straight line increases the singularity by the triangle will also bolster giving equal potency in singularity by the half circle. As the singularity of the major component revives the lesser singularity to equality, the triangle in singularity will match the performance and so would the half circle respond in precise ratio setting equilibrium in order.

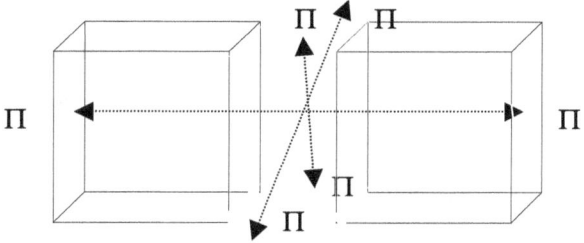

From this the lesser partner will fill by the extent of the larger partner and as soon as equilibrium sets in the growth will duplex in both accounts, normally to the fatality of the lesser partner as the lesser partner will not be capitulating under the straight of the duo. The Titius Bode configuration in accordance to orbiting formation holds a slightly different explanation to the explanation that applies to cosmic structure surrounded by space. It is moreover the individual singularity in maintaining the major singularity, which sustains the governing singularity providing equilibrium in space-time.

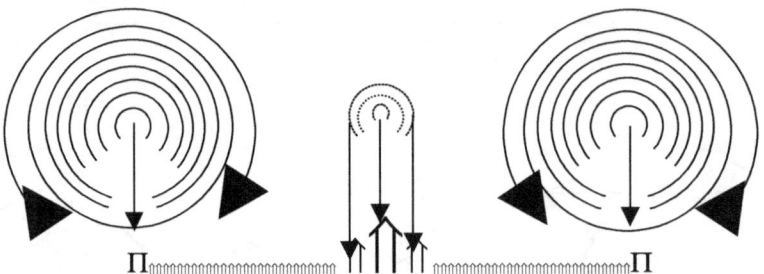

Not only does atomic individual singularity maintain self preservation, but in doing that it also sustain a governing singularity holding structural composition and form within a cluster of matter for example a star. Between stars there are a mutual or bonding singularity between atoms and stars.

The sectors provide individual singularity a means in sustaining governing singularity by which provision comes through maintaining governing singularity the required spin in maintaining cooling. If this process did not apply, there would be no connecting individual singularity to major singularity. The sectors provide individual singularity a means in sustaining governing singularity by which provision comes through maintaining governing singularity the required spin in maintaining cooling. If this process did not apply, there would be no connecting individual singularity to major singularity In this maintaining of cross referencing of singularity providing spin to the governing singularity many factors of singularity all form a close knit network inseparable one unity but also strictly individual to a point of destructing.

Singularity has three part and five points with Π as matter being sixth and space (r) as light the seventh.

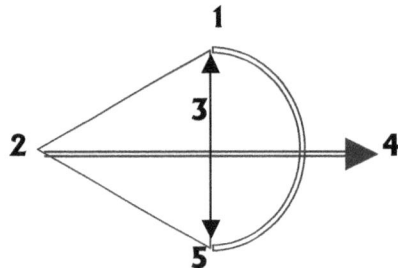

The TITIUS BODE Principle

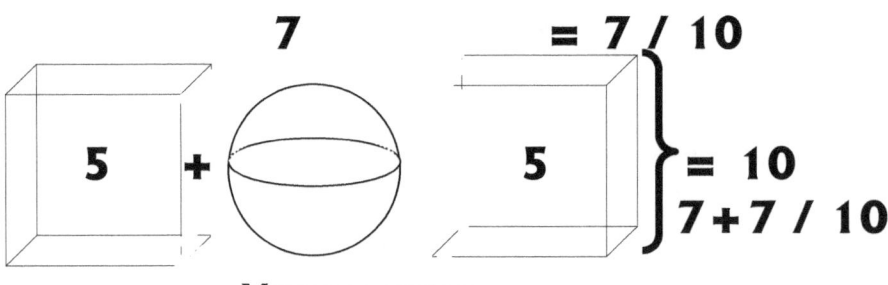

Matter-to-matter

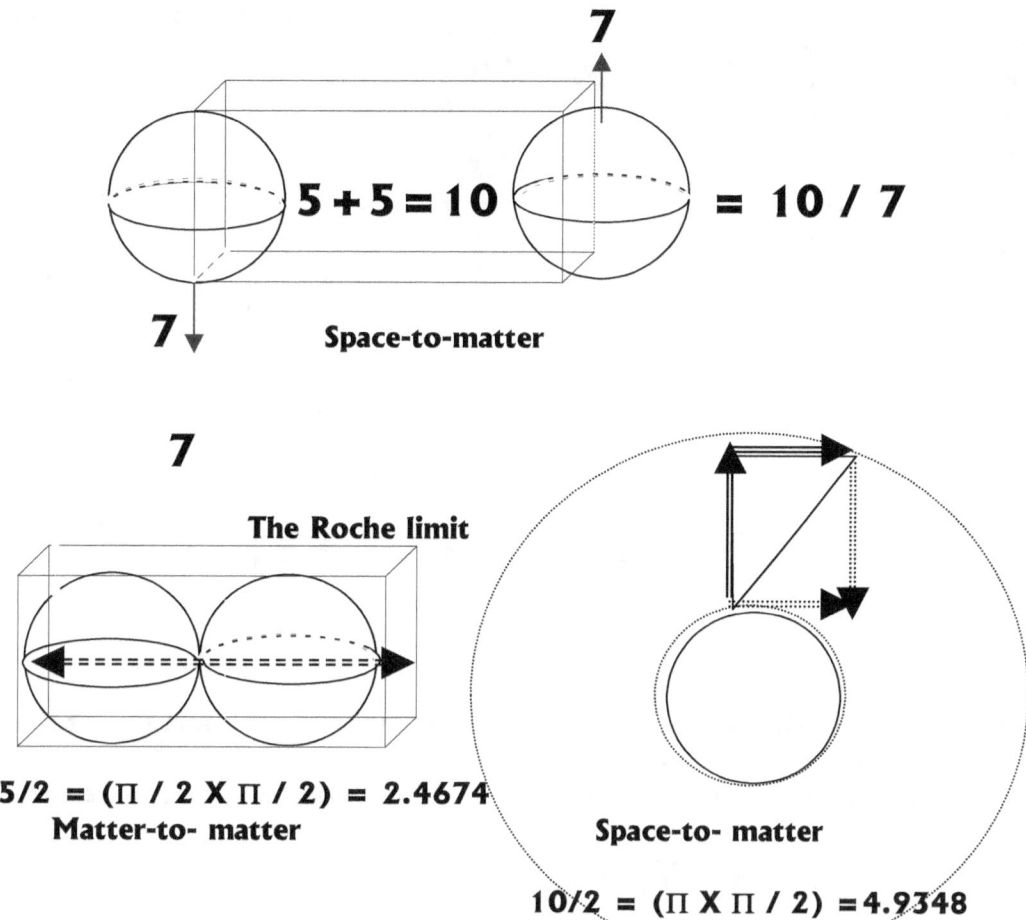

The space between the spheres divide in half, but because of the extending of Π and not applying r as ordinary mathematics will suggest where Π replaces r the singularity extending from $Π^0$ will be half of Π in the square of $Π = (Π/2)^2 = 2.4674$. In this lies the dynamics why planets have a positional (be it rather a dimensional) relation of 7 / 10 The second Roche limit is within the sphere as $(Π^2/2) = 4.9348$.

In this gravity or movement of the earth holds one Π and the capsule coming towards the earth holds half of the value of movement caused but the other Π or gravity mark connected to the movement of the earth. This places the total value of movement at (earth) Π x (capsule) $Π / 2 = Π^2 /2$ and the numerical value of $Π^2 /2 = 4.9348$.

Planet	Mass per Earth unit	k^{-1} Movement	a^3 of space volume	T^2 During time units
Mercury	0.06	$T^2 \div a^3 =$ 0.983	$(a^3)=$ 0.059	$(T^2)=$ 0.058
Venus	0.82	$T^2 \div a^3 =$ 0.992	$(a^3)=$ 0.381	$(T^2)=$ 0.378
Earth	1.000	$T^2 \div a^3 =$ 1.000	$(a^3)=$ 1.000	$(T^2)=$ 1.000
Mars	0.11	$T^2 \div a^3 =$ 1.000	$(a^3)=$ 3.54	$(T^2)=$ 3.54
Jupiter	317.89	$T^2 \div a^3 =$ 1.000	$(a^3)=$ 140.6	$(T^2)=$ 140.66
Saturn	95.17	$T^2 \div a^3 =$ 0.999	$(a^3)=$ 868.25	$(T^2)=$ 67.9
Uranus	14.53	$T^2 \div a^3 =$ 1.000	$(a^3)=$ 7067	$(T^2)=$ 7069
Neptune	17.14	$T^2 \div a^3 =$ 0.999	$(a^3)=$ 27189	$(T^2)=$ 27159
Pluto	0.0025	$T^2 \div a^3 =$ 1.004	$(a^3)=$ 61443	$(T^2)=$ 61703

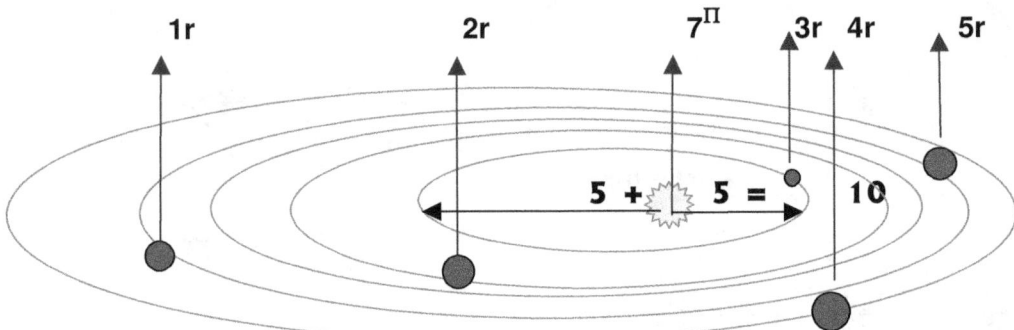

From the matter-to-matter relation in the Titius Bode configuration there are 7 / 10 + 7 / 10 = .7 + .7 = 1.4

From the space-to-matter relation in the Titius Bode configuration there is 10 / 7 = 1.42

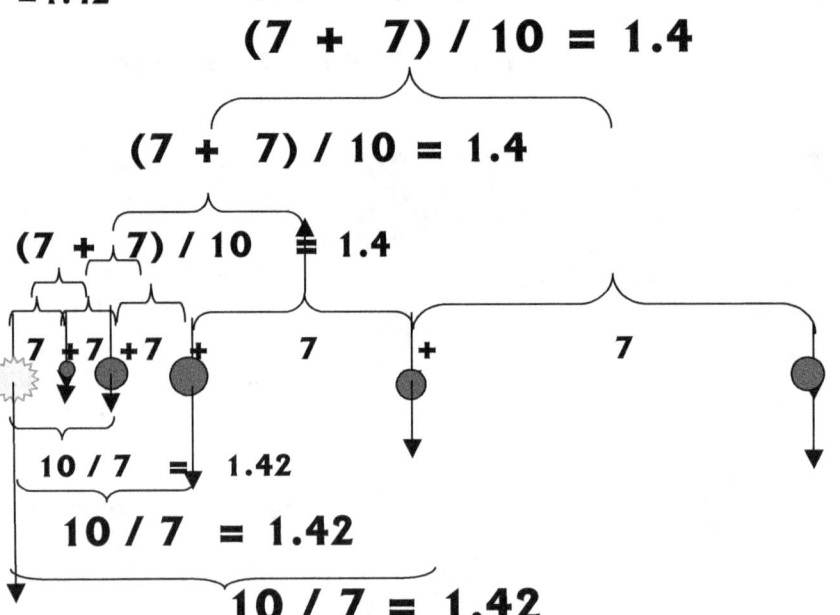

= .7 /\\ 1.42

= 1.4 /\\ 1.42 Because the space-to-matter is in the square at 10 placing the matter-to-matter at a square of .7 + .7 = 1.4 the space-to-matter forces the matter-to-matter to double the distance by number as structures are place father from the mainΠ^0 maintaining singularity.

Reasons why this does not fully apply to the solar system I give in book # 7.

According to me, as a believer of the Holy Bible, there are seven heavens. The Holy Creator sits in the seventh heaven. His Word tells me that. That means if his Word says there are seven heaves that means there are seven heavens, that means there must be six undetected heavens. Before those atheists start shouting their lungs out, first go about answering the question put by me in a previous article. Answer the question where do life, being energy, go after it has left the body after death. Let us presume that the question is still without an answer. If the Creator lives in the seventh heaven, then He must have created the other six. By acknowledging the creation, I echo what millions of intellectual people believe. The so-called Christians, Judaism and Islamic faiths all accept the first five parts of the same religious Bible. Therefore, I feel free to speak on behalf of millions of people who consider themselves religious and these people come from all lifestyles and all over the world.

First let us consider the dimensions accepted by science and later argue those forms that are excluded. The so-called heavens have another name, which are dimensions. These dimensions could be regarded as planes or spheres. In looking for the dimensions that form the universe, one must look for sides, which has a combined value, but exists in total isolation from one another. Any object that can be visualized has to contain at least three sides with six obvious different spheres. These spheres do have single applications in the universe. Allow me to explain.

THE SINGLE OR FIRST DIMENSION (GRAVITY)

In the single dimension, one finds gravity or a pulling force that every cosmic body has.

If any person does not believe in gravity, then try to jump high of far. One would find that there is a force, which pulls your body back in the direction of the centre of the earth. This force is calculated to be 9,81 N m/s^2 and is almost precise the same value right over the world. This force can only be overcome if an object is hurled into space at a speed of 11,20 km/s^2 and an angle of 90° with the earth. That brings about that any object on earth is moving at a speed of 11,17 km/s^2 at any given moment.

As far as my knowledge goes, this force is perpendicular to the earth, with a distortion of 7% due to the inclination of the earth. A person in China will be pulled in an opposite direction to a person in America, because China and America are approximately in opposite directions to one another on the face of the globe. This means a person in America moves towards the earth at 11,17 km/s^2 which is directly opposite in direction to the 11,17 km/s^2 which the person in China is drawn to America. Taken these viewpoints into consideration, gravity is a single directional force moving only in one direction that depends on the body position to the earth. Where does this force stop? Nobody knows, because the crust of the earth is being drawn towards the centre of the earth. Even the see and the crust underneath it moves statically towards the centre of the earth at an even pace. If one looks at this force, one gets the impression that there is only one direction towards the centre of the earth without any given point where it will stop. On the other side there is no given any definite starting point. Without a star or an end, only a direction remains that envelops the whole idea.

I hope I was clear enough about the idea I tried to explain. The whole idea of gravity consists of only one single directional movement in one direction seen from all directions. There is no starting point, no point where it ends, only the definite direction that the body moves towards the centre of the earth relative to a point where all directions are measured. Another strong candidate of this dimension is magnetism. The polarity of the atom in the iron core is only towards one direction. But I shall explain a little further down the road, why magnetism do not comply with the single dimension force when I have brought some more facts and arguments in explaining what forces come into play.

THE SECOND DIMENSION : THE WAVE

In the context of the wave, for example water, the circle of the wave flows from one given point outwards. That implies that there is a definite direction. However, there is a second dimension in the wavelength. The moment that the breadth of the circle is determined, the wave has already altered it. That means there are only two values to be considered realistically without freezing time and that is the direction and the frequency. Sound and light are two such values. The moment the wave is frozen in time, which means it is standing still, it is no longer a wave, but an unnatural structure man made for his own benefit. It does not occur in nature because a wave can never stand still.

Light, sound and waves are two-dimensional values. The dimension consists of a point of origin and a frequency. There is no distance, because the wave beams out in all directions simultaneously. The Doppler effect in sound has an influence but that is because the point of origin keeps altering due to the movement of the source. The speed of sound has a definite ratio to the density of the medium it moves through and that applies to the time value of the space-time occupied by matter. In a later part, I shall explain this in detail.

Due to the changes in space time occupation of the matter the frequency alters and tarnish directly in ration to the space and the time the matter occupies. As the frequency of the space deteriorates in time, the wave becomes less in value but still remains. Therefore, all the sounds the dinosaurs made are still with us but the space-time occupation that the sound waves had, rendered it undetectable to us. Proof of my argument is found in the heat that still can be detected with the Big Bang event. Although the space-time value is only round about 3 °K, the waves can still be measured, because these waves form part of space-time. The light wave is transmitted from its source in a sphere and will transfer itself through space-time until such time that a few particles hits a solid object. The wave consists of billions of light particles that move by wave density from the transmitter. The photon moves at a rate of ± 300 000 km/so. That means that the photon displaces space-time by negatively at the above-mentioned rate. This negative displacement of space-time by photons is called a light beam.

The light rays follow wave upon wave in endless motion through space in time. As soon as one wave of photons hit solid matter, only a very small number of the total wave is stopped. These meaningless stopping causes a shadow to form, but the large number of photons available would fill the gap formed by the loss of photons. That means that a shadow as large as the earth, will cast darkness for a very small distance / time span that means a tiny part in space-time.

The reason why the heaven does not light up is that the beam is scattered as it spreads out in a balloon formation and therefore the number of photons lessens in density as it becomes "duller". The more space-time it has to displace the thinner the layer of photons would become and the less intense the beam would seem. It is not so much the space (distance) that should be regarded but the time it spends in space has the ultimate segregation of the light beam. In space and time, the beam would become so minute that it would alternately become unobserved by the human eye.

All that we know, all the knowledge we have accumulated through time is based on the light that sends information to us. The size, the distance, the structure, the density and its position are all determined from the light that sends information to us. The light that reaches our planet determines all the size, the distance, the structure, the density and its position. All our facts are based on this small amount of evidence. Light is so widely connected with insight and knowledge that we assume that we have seen the light when a picture of a person is projected with a light bulb next to the person's head. Light is seen as the same as knowledge. Light is actually a very poor source of information. Magicians and con artists use the information of light reflected by mirrors, lenses and colours to mesmerize our wits.

Let us consider light as a source of information in daily use by humans on this planet. Light is a very poor medium of information, due to all kinds of illusions, which it causes. Take for instance how mirrors and lenses can disturb a person's image. Why then can we categorically and without doubt, be assured that the information that we collect are the truth. Still, those in power disregard any phenomenon they find as the truth. A very practical example to prove my point can be put as follows. Light shines on a green leaf. The leaf is reflecting a certain variation of light and the rest is absorbed. Let us picture the leaf olive green. The olive green leaf accepts all the colours in the spectrum, except that olive green. The olive green are not accepted by the leaf but are rejected. The part of the light spectrum that seems to be unacceptable to the leaf is the very colour we associate it with. If the leaf accepts all the colours except green, it must consist of all the colours except olive green. Another example we use every day, is "How bright the moon shines", when in fact, everybody knows that the moon does not shine !

This might seem trivial and irrelevant, but remember, because of such trivial terminology, people were burnt on stakes. When Georigiano Bruno tried to persuade the church it was not the sun rising and setting, this became his fate.

Man's attachment to his visual sense is actually a little funny and a lot deplorable. Science can calculate the density of a neutron star, they can determine the heat value in such a star, but when this star becomes more dens, to such an extent that the density causes the neutron star's atoms to disintegrate, these very same scientists declare that:

1. the star has vanished;
2. it has gone through to another universe
3. it does not exist any more
4. because it is dark, it must be cold
5. it is lost in the creation, never to be recovered again
6. with this, all other laws of science are disregarded that says an atom consists of frozen energy;
7. the gravity fields can still be mesmerized therefore the matter it consists of must still be there;
8. All that matter can still be placed at one certain, predetermined place, and therefore it is position is a fact as far as the universe is concerned.

When taking the above mentally into account, one cannot but wonder, how far did man actually progressed from those darken middle age mentalities.

With these very obvious facts in the fore ground of our minds, we still reassure all the doubtful thoughts of our ancestry, which regarded everything the unknown comprised of as magic and mysterious.

We still cannot appreciate anything that falls outside the boundaries of the visual senses that the instruments can detect and determine. How small we are and how inflated we regard us to be.

I have in the past been asked the question "Why are all these other stars and galaxies necessary , if we as humans can't use it?" The human arrogance has no limits and the common position man occupies in the micro as well as macro space, is completely distorted by our sense of self-appreciation and self-importance.

THE THIRD DIMENSION (THE ATOM AND MAGNETISM)

In the article that deals with the meaning about "nothing" the structure and lay out of the atom is explained to some extent. The atom is the smallest and the largest single unit that the universe comprises of. In the most basic form, the atom exists of one single electron that orbits to energy levels around a nucleus in a single electron lay out. This electron to nucleus balance is the precise force that keeps the universe in perpetual notion bound by time. In a later stage, I shall explain this in more detail. The energy levels that the electron positions itself to the nucleus are valued in quantum leaps. However, even this complies with a three dimensional length, width and depth. This means there are three definite measurements, although they seem to be microscopically.

Taken one step further, the atom is made up of frozen energy. Any matter that moves at the speed of light is pure energy ($E = MC^2$) according to Prof. A. Einstein. That means this atom cannot be changed in shape or size without enormous energy loss or gain and this would lead to extremely serious consequences. The Japanese at Hiroshima and Nagasaki can declare what extreme consequences are hidden underneath the structural disfiguration of the atom structures. Therefore, can the people in the Tunguska river valley give evidence to the outcome of atoms that gain in mass.

In this process, man has released the worst kind of destruction available. The energy released is of such vast dimensions those 50 years after the explosion the shadows of the victims are still edged out on the background in the cement and bricks. The intensity of the light that was released in a billionth of a second, has almost forever changed the face of the material it shone on. Spare a thought for the people who received that light onto their bodies as their shadows remain as testament to those in power who damned them forever.

This process was not caused by atoms that was demolished, but merely by changing some of the element value from one atom to another. This leads to a spontaneous thought: "Why has no American ever been brought to justice for horrendous acts of war crimes?" Needless to say this act was the mother of all war inflicted war crimes by killing and maiming hundreds of thousands woman and children and civilians. These bombs made those in power, who ordered the release of these bombs, the biggest sadists and mass murderers ever known to man and when compared to evil-minded monsters like Nero, Nero with all his menace, suddenly becomes a silly and naughty boy. Even decades after the release of these highly toxic energy sources, they still kill and maim the innocent. However, this only ties in with another admirable fact of the 20th century. Almost all nations on earth have produced war criminals and warmongers, except the English speaking nations on earth. This group is blessed with the innocence to such an extreme that not one has ever been charged with one single act of a war crime. If none has been charged, none could be found guilty and then not one can be guilty of war misconduct. Let us go back to the atom's structure. Because the atom moves about at the speed of light, the structure is timeless. All matter that is timeless, will last forever or eternally.

This brings about the atom's structure to be forever and timeless forming the third dimension. In a later part, the reader will find that I disagree totally with professor Einstein's assumption about the fact that the speed of light has the same value as time itself. In order not to start confusion this early in the book, we shall accept dr. Einstein's theory as for now. The message I am trying to bring across in this article is the comprehension of what the third dimension contains. It is made up of three dimensions without having time as a factor.

MAGNETISM: THE MISCONCEPTION OF THE ATOM

As I already pointed out, the atom is pure energy. Seeing that the electrons rotates about the nucleus at the speed of light, and the nucleus vibrates (?) at the speed of light, that brings about a confined cell made up of pure energy.

Seen in the whole picture, the fact that the atom is driven at the speed of light and is in total balance and harmony, its permanence lies in the fact that it exists confined to a structure, but not to time itself.

There are length, width and height. The tree dimensions it composes of will render the structure eternal life, or that is how it is regarded by science. The property of the atom might change from star to star, but the structure remains with its three dimensional qualities varying in size, but it remains the same. You, the reader may ask: "What is the common factor between the atom and magnetism?"

Magnetism on the other hand flows between two points without stop in a closed circuit, not influenced by time. There is a direction (length), a circuit (height) and a start / finish point. This forms a closed ring formation, which has the same qualities as the atom. Time plays no role in this energy displacement, because this movement is coupled to the speed of light.

To prove its existence, lies in the fact that the human civilization is mostly driven by electricity.

The energy determines the magnetization, but the circuit remains permanent although some times in a latent form. Because of this, it actually consists of all the ingredients to qualify as a third dimensional force.

The ferromagnetic field proves that space-time is being displaced as it is being done in the case of the atom structure, but the displacing of the space-time is in a continuous and constant closed circuit.

The poles that attract each other displace space-time in the same direction and those that repulse each other, displace space-time in opposite directions.

In bodies as insignificant as the earth, the difference between gravity and magnetism will be enormous in comparison with structures the size of the sun. The magnetism in the sun is comparatively much stronger than the earth, but the difference between gravity and magnetism would be less. In a structure, the size of a White Dwarf the electromagnetism might only be twice the force of gravity and accordingly as the star becomes bigger, the force would become equal.

In structures that compose of the mass of a neutron star, magnetism would be dominated by gravity to such an extent that the force of gravity would not allow any magnetism to exist. In the so-called black hole, there can be no such a thing as magnetism because the electron does not exist any more.

Electromagnetism is the "short circuit" in the flow of positive space-time displacement and can only exist in a structure that compiles of an iron-based core.

THE FOURTH DIMENSION: THE TIME FACTOR OF TIME IN SPACE

The previous three dimensions all dealt with the space factor in time in space, disregarding the time factor. Let us consider an everyday household item in normal use like a table to explain the value of time. The table is made of wood. The wood is comprised of timeless atoms. (That is, if the reader accepts the previous argument about the atom that lies outside the boundaries of time). The form and the shape that the wood is in, is not timeless. It started as a seed that was enlarged by cell multiplication as it germinated to become a tree. The process of germination took a certain time and the growing of the tree took another period in time. That means the tree occupied more and more space in a given time period in the space-time it shared with the earth in the form of a tree.

Afterwards the tree was chopped off. This felling of the tree comprised of a certain given space in a certain given time, as did the falling of the tree. Both these periods consisted of different space in different times that are coupled to the space-time the earth occupied. For a certain period, the tree remained in an upright position occupying a little more space as time moved on. Then at a predetermined point in time, the tree was felled. Every blow by die axe displaced a certain piece of wood to a different place in space and time. That meant that every splinter of wood that broke from the tree was given its own place to occupy space in time. The position of the wood has been altered and even if one try as hard as they may, every piece of wood has received its own space in time to occupy and could never be regarded as a tree again. It occupies a complete different position in the space it shares with the earth in time.
Afterwards this tree is stripped of its branches and leaves. The stripping takes a certain period and position in space and time. Each branch occupies a different position in space and time and is forever dispositional from its original position in space and time, because every part that is not part of the tree anymore, is in its own position in space and time.

Then this, the tree was taken from the plantation to the mill and moved through a different space in each fraction of time as it was transported. Every millisecond held a different position in space in relation to the next time fraction. At the mill, it was sawed into planks and the structure's position was altered even more in space and time. The planks were bought by a carpenter and were given a completely new form as a table in space and time.

Although this particular table I am writing on now, at this present moment, has occupied space in time since 1921 as a table, the wood occupied space in time much longer than the table has in its present state. This wood can never be made a tree again, because it is space and time has been altered indefinitely. So, how does this fit into the big picture of the universe as all things in the universe are connected and related?

I was born on a given second, minute and hour and I shall die on a certain second, minute and hour. From the day of my birth until the day of my death, I shall constantly alter my position in space and time and will never be able to occupy the very same spot of space-time because space and time will not allow it. According to my visual observation, I can presume to occupy the same space in time, but the geodesic outlay of the universe is altered as every millisecond goes by. That means I can never remain in the same space in time, because my body's position alters by the rotation of the earth, the sun and the universe.

When my time is up, my space and time is altered to such a position where as I am no longer in control of it. From then on, I will never be able to determine my own position in space and time again. I am in a state called death. In this state my body will be broken up by microbes into gas and heat, every second time moves on, my body's occupation of space, and time would diminish. This will carry on in space and time until only the elements my body comprises of, remains. The atoms I am made of, have previously been used to form plants, trees, animals and even humans. After my death, it will never ever form another combined unit to match my exact replica. Therefore, I have departed in more senses than one.

That means my position as Peet Schutte, in space, and in time, is suspended, unconditionally, forever.

This means that for certain duration of time a certain combination of atoms is forced together to form the elements that are dedicated to me. This dedication is temporary, which means that I was for a certain designated space in certain duration of time-sharing space-time with the earth.

Although this is a known fact to every person on earth, it is astonishing to see how every person yearns to maintain a youthful and vigorous maternal structure. This structure is condemned to destruct the minute a person is born, yet everybody guards his attachment to that structure with a jealous observation. The cosmetic industry cashes billions of dollars a year and all that income is based on this fear of ageing, which ultimately leads to death of the person and destruction of the body. The ongoing process in the envelope of the fourth dimension connects all the previous three dimensions, we regard as time. This means that time itself is not created by man, but is part of the physical universe and is only recognized by man, the very same way man has recognized the existence of the other three dimensions.

THE FIFTH DIMENSION: UNOBSERVABLE - THAT MEANS TIMELESS TO MAN'S SENSES

In the article "Life after death," I touched on the subject that life is a form of energy, and therefore is indestructible cannot be destroyed. If any reader cannot accept this argument, then please prove the opposite and let me know. Therefore, until the opposite is proven, I shall regard this argument to be correct. Because the generator, which I regard to be me, I of the electrons, is not the" me" that will be placed in a box and will never be able to share space time (because the worms are going to feast on me) I am above time and space and I shall not be in that coffin. Only the "me" that am made of flesh and bones and was considered "me" and which I had temporarily control over, will deteriorate in that coffin. If the generator of the electrons that is somewhere connected to the brain, and from where I control the me, which is the muscle bone and tissue held in place by the body I consider to be me, where do I , the generator of electrons go after I was disconnected from the me of flesh and bone me. I know where I, the energy-less dead body is going. I, the part I consider me, is to be thrown into a wooden box, dumped into a hole which is dug in the ground and where the worms are going to enjoy a feast of a meal.

This feast of a meal by the worms cannot be me, because I am not part of that decomposing structure. I was part of that decomposing structure, until the very second, I lost control of that decomposing structure. Thoughts, that travels faster than the speed of light is the actual part that is I. I can elaborate on this line of argument, but those that do not wish to be convinced will remain so because they prefer to remain unconvinced, not because they know they are right, but just the opposite. The I which is regarded as me, has received the ability, for a short while at least, to manipulate space-time, whether it was in the form of my body, or other matter I came into contact with, or unoccupied space-time as I moved about, occupying unoccupied space-time in a random fashion at free will.

The part that is I, and which is not part of the corpse any longer, can think of one thing and then think of something completely else. The very following second I can change my surroundings as soon as I change my thoughts by creating new thoughts. These thoughts are not connected to time or space. It moves arbitrary and involuntary through time and space, from present to past to future. Sometimes these thoughts are so strong that a person loses track of reality.

The terminology we use to describe this, is daydreaming. When I was teaching as a pedagogue in class, this condition was my biggest enemy. While I was conducting my class, the students would sit there wide eyed, listening, but at the same time they were miles away from school living a daydream that had no connection to matters of schooling. When they were asked to reply what was just said in class, they were flabbergasted and completely unaware of their surroundings.

I think I may presume with some certainty that the reader would follow the two parts of the same person I meant. I agree that an argument can be made that these thoughts are part of brain cells that are stored by nerve tissue, but the information in those cells are created by emotion. These thoughts can be depressed by emotion or be prominent on the foreground because of emotion. Man is made up of a stream of emotion that flows continuously through these emotion fields and the flow of emotions is that that puts meaning into a person's life. One has to consider emotion to be one of the most pure sources of energy.

Energy creates thought, creates ideas of a spiritual as well as physical nature, it establishes a flow of electrons that drives the human mind and body, it controls all muscle groups in the body like heartbeat and when the body is in danger, it produces chemicals which enables the body to react far better than it normally does. That means these other organs are also under the control of emotion. The emotion drives the body to produce chemical substances, which enables the body to perform at levels far above its known abilities. This emotional control defines heroes from cowards, sportsman from the ordinary and even philosophers from the masses. That means the emotional part is the part of me that cannot be destroyed and that is the part that generates emotion. All this comes down to one value that life possesses, and that is the manipulation and control of space-time.

When this emotion driver, that generates electrons, leaves the body, a person is considered dead. When one takes the moment that life leaves the body, nothing physical leaves the body. There is no lightning like electric conduction, there is no immediate spontaneous combustion, but there is no emotion either. As I already pointed out, the only reason why a person is considered dead is that he is considered lifeless, meaning without energy.

A lot of energy has to be deplaned somewhere. It is no longer part of the physical world and being energy it cannot just vanish into nothing. What is factual, is that life as energy, is no longer attached or bound to the fourth dimension of time. At this point, I have to prevent my ego and self-importance not to get the better of me and put myself on a pedestal equal to our Creator.

It would be much better if I were grateful and thankful for the time I was allowed to use the atoms loaned to me for my own personal use. The loan period was of such a short duration in space and time, that the extent of the period of loan becomes oblivious in space and time and space time.

THE SIXTH DIMENSION: THE LIGHT AND THE TRUTH

I must confess that in this my faith and religion plays a large role, as one of my beliefs is that the Messiah has already come.

Maybe the Jews, Moslems, Hindus and other religious groupings would find their own explanation according to their faith. My Messiah said, "In the house of my Father there are many mansions.", which I interpret that these groups should be left to find their own salvation. I may not condemn them or denounce them or try and convert them, because each should have its own mansion. However, personally I accept my Messiah and He declared that: "I am the Light and the Truth and no one can enter the house of My Father but through "Me."

If my Messiah points out to be a light, it stands to reason that there must be darkness. That will be a dark fifth dimension and light sixth dimension. Seeing that the majority of English speaking persons do not share my beliefs and religion, and I do not believe in converting any person to my faith, I shall leave this matter at this point.

It is not that I am ashamed of the religious beliefs that I follow; to the contrary, I believe that only Israelites may be converted to my religion. I think I made enough argument to prove the existence of a fifth dimension to which life has to go after death. I presume, because I am not familiar with the contents of other religions, that their religion will allow them a passage by what means their religion chooses, out of the dark fifth dimension.

This dimension, the sixth can only is entered by human life. Let us then see what human life is all about. A human does not have to be a human because the creature is compiled of human D.N.A. D.N.A. cannot form a human. The gorilla is 97,8% human and the orang-utan is 98,2% human. However, there are obviously big differences between these species and a human being. No cultured person on earth will consider one of the ape species to be human. That means a human is more human by culture than by physical appearance. The additional 1,8% that a human have, is not even enough to explain the physical differences that exists between humans and apes. However, we know that the main difference between the species is the human's ability to reason, think, argue and control their emotions and instincts.

The more a human explore and scrutinize his feelings, himself, his surroundings and his universe, the more such a person would qualify to be human. After all, that is how other species evolved away from the animal and that should categorize us to be part of the sixth dimension.

THE SEVENTH DIMENSION: THE SUPREME ALMIGHTY
Because so few English-speaking persons consider themselves Israelites, they do not fall under the law of the Israelites. In such a case, I would consider myself blasphemous to share such knowledge with those that do not regard themselves to be law-abiding in all ways. Any person that does want to know more about this matter should read the book of Henog. I will say this much, that those that read Henog would find out why the book of Henog has been left out of the Holy Bible by the Roman Catholic Church and the other churches. The Supreme Being lives in the Seventh Heaven as Creator of all. Because I regard my fellow Boere as brothers, I did explain to a very small extent the seventh heaven in the original Afrikaans version. But all Christians and only Christians read this…all churches are part of the Anti-Christ being the Body –Of- The- Anti-Christ. Do not look for the coming of the Anti-Christ for he is among us. They crucified Christ for His throwing over the money tables and throwing out the money offerings (a lucrative business in any society) ridding the Temple of money some two thousand years back, and today all Christian religions fight one another to feed that which Christ threw out…the money tables…and best of all is they feel righteous in doing so! All denominations are a part of the Anti-Christ and BEING THE ANTI CHRIST. Christ threw out the money tables because He said you cannot serve two Masters…you cannot love God and Mammon for one you shall love and one you shall hate. You cannot serve Mammon and God. If Christ came back today and once again threw out the money tables all Churches and worshiping priests will once more shout for His crucifixion as they did two thousand years ago. As the Pharisees were the Anti –Christ back when… so is all Christian churches and denominators, Priests, Pastors, Reverends, Bishops, name them what you like, small and large…they're all taking part in the crucifixion every time they ask for money "In the Name Of The Lord" and if He came to destabilize they're Money machinery today, they will hang him tomorrow morning at they're earliest convenience and even on the same cross (if they can find it). Every preacher is more into collecting than pouring out the Word. It is a trade off that the preacher will bring the Gospel in exchange for collecting offerings. They heel, bless, pray, condemn and condone on behalf of the name Mammon. I challenge every purist of heart to show me one preacher of the Gospel that sends donations back with the message that such donations is condoning the Crucifixion and as preacher will not except having the blood of

Christ on his conscience. If you cannot show me one, I can show the body of the Anti-Christ for they will kill again if some one should try to diminish the lucrative trading done in the Name of the Lord.

9.6 THE TRINITY

As far as the aspects concerning the physics side goes I could prove in some way that my initial way of reasoning brought fruit to bear, but what about the dimensions past four and even past five. I think I may presume that in some way I did manage to show that the body has its place in the fourth dimension and without life it becomes just one more cosmic structure without any form of moving ability when all forms of life (including the bacteria that will decompose it to atoms) removes from the carbon and other elements where the elements then are exclusively and only cosmic particles. With good reason one may believe that the ability in conducting the manipulation of the space-time within the body has diminished to all extent and none is any longer present. To find a means of putting mathematical formulas to use in applying proof to indicate the fifth dimension is beyond me and that is where human extelegence plays the part. By the same token that atheists can say they wish for more proof about life and the fifth dimension I can demand the explanation to prove otherwise and ask an explanation about the energy presence in life-holding bodies and the energy absence in lifeless bodies and where as I doubt I may force atheists to sound understanding they too must admit that something does seem out of place in they're arguments about the energy being of a pure physical nature and only stubbornness will win by the days end.

So to them and those, as I might not find a way to prove beyond doubt them and those also must admit I have sown some doubt in their minds. Getting them to admit with some gallantry about the doubt factor, well that I must gallantly admit is a horse of another colour. With the fifth dimension seemingly impossible to prove mathematically the sixth will be much more difficult to prove and I shall not even attempt such an act. Fortunately man is not mathematics but much more complicated than mathematical equations can ever bring forth. Only good old fashion arguing with a dash of logic sprinkled when and where necessary will pave the way to understanding man and beast. To find out what man regard as good or bad and what beast evidently regard as good or bad must be different as everything else about man and beast are different and we must scrutinize beyond where mathematics and physics can prove. Even the most convinced Newtonian should see that there is a point such as that.

Being human every reader must have an opinion about good and bad and what is evil and what is not. I can scarcely imagine a lioness feeling bad about a kill while her cubs are filling their bellies with mouth-watering flesh of an antelope kill. Neither have I detected sorrow and anger as a new male kills the previous litter to establish his new domain. Neither the female nor the male bears sorrow after the deed of him destroying her litter sired by a previous dominant male although the lioness will protect her cubs while they are alive almost to the point where she may put her own life on the line. That how ever has nothing to do with rite or wrong good or evil and after the lion male did his killing of her cubs she shows no remorse or blame for that matter as she follows him back to the rest of the waiting pride. This is not exclusive to lions or even to predators but has a wide range of animals following the very same living style. Horses kill without thought because when a stallion wants to find a mare that he knows may follow him but for the foal by her side he will kill the unwanted foal and not be bothered by her reaction, being in the knowledge she will follow him after the foal is dead in any event. While the foal is still alive she may do some protecting of the foal but she will normally not go as far as the lioness in preventing the death of the foal. Baboons, monkeys and a variety of animals have this approach to life. During such an attack by the new dominant male baboon the females will fight off the onslaught either by grouping together or by fighting him off in ones and twos but the usual is that the new male is big young and strong and even a group of females are no match in a fight. However after all the noise and the shouting blood sweat and the rage of adrenalin has died down, the babies are dead and no female attacks the male after the fact in heart felt sorrow for the loss they feel such remorse over. No, it is clear that the deed was done and life goes on.

In humans such behaviour does take place and every time we read about such a deed committed against a harmless infant even the biggest humanist find a moment where he or she wishes the death penalty on the criminal for acting in such brutality. Why will humans shout for blood in punishment while other species take it in stride?

To find a solution I do what I always do. I turn to the Bible for an answer. (You atheists deny yourself a wonderful encyclopaedia of information and when reading the last book in this Theses you will come to understand my saying so.) In The Theory which is the seventh part of The Theses I explain to very detail the events of the first six days of creation as recorded by The Authentic Biblical Author but even after that some more explaining arrive when it is correctly translated. In the book called the Bible there are two trees described with distinction and I am of no opinion to whether they were in wooden fibre or just symbols to explain the complicated issues to persons with even lesser education back then than I have at present. It is an undeniable fact that man has an inclination about what is good and what is evil. Man would not kill an infant because he cannot find the infant's mother and the infant is crying of hunger. In the animal world any adult of the specie will walk past such an infant in distress with no feelings of care what so ever even if the specie does not show a normal tendency to destroy such an uncared suckling. Where one mail may kill another mail in a fight to establish dominance the group does not cry for justice as they loath such a deed. When a superior member of a pack relieve a lesser member of food or eating rank they do not hold congress in judgement to provide sanction for the lesser member with accompanying reprimanding about such incorrect behaviour by members of the group.

Such is the caring of man that any human notwithstanding whatever urgency drives him at that particular moment that human with maternal instincts or not will stop to care when coming across a deserted human infant hungry and all alone. Why will man show such behaviour as normal through out all races on earth without any cultural distinction in any way and this may be the only distinction that races share because some eat their dead and some burry them with pity filled emotion and... Oh, I can go on writing a book on this topic alone but that will be useless because every one should know what I mean. With this shared by all where and what does mention this distinguishing behaviour of man's ability in judging between good and bad for the very first time as a landmark to man.

Accept the Bible or not, it remains the oldest Book available on matters reported by man since no one knows when because Moses may have assembled the research on information but the information as such dates to times predating even what Moses' research may indicate. In addition it may be correctly presumed that many or most of the facts he recorded he was taught as a prince in the house of pharaoh and that then may explain some detail about matters he actually knew nothing of. His being adopted by pharaoh's sister must be a plan with some significance and must result in some meaning other than to give Moses a childhood of luxury alone.

One of the trees the Bible mention carries a name specifically as life and from that I draw a conclusion that may reflect that life chose to go the way of having a variety as sells with complexity in the evidence we now gather from DNA strands whereas a choice of life would indicate a sell of simplicity in structure as we find with lowly developed insects and other specimen of life. I have seen how a corn-crake of a specific kind only found in the desert and semi desert regions where I live can start a pest becoming so out of control that when run over on the road they form layers of millimetres thick trampled and squashed by cars to the extent the tar on the road is no longer visible. It truly is a pest of Biblical proportions but fortunately to unleash this pest it has to rain in November in one specific week. If it does not rain at that specific date, and I do not mean approximately but precisely they don't show at all. That makes their return very sporadic and it only occurred about five times where two of the five times became a pest like none can repeat in the more than twenty years I farmed on that farm. The pattern also follow a distinction where at first one or two may show very sporadic in places. Then mating starts and the pest develop where by it truly come to a climax at the end of February and dies down in May. From a few eggs they develop millions on millions and I do not exaggerate in the least when I refer to this phenomena as a plague of Biblical proportions. Poison does not kill them and when one gets hold of another one the bigger one just start feeding on the smaller one. At the end of the meal where the bigger one devoured the smaller one in totality (and I mean boots and all) the specimen will shed its skin immediately, eat it or not eat it and walk off. The one is a precise duplicate of the other with no distinction amongst them of whatever nature. Seeing one is seeing the lot. They share genes in precise replica with no differentiation of any kind what so ever whereby the one may have even the tiniest of difference in any form. They come from a line where life was still very basic and the mother specie of the very original has not changed in any way through millions and possibly billions of years. That I say on the grounds that to my judgement that specie is of a very basic nature and has developed with one single aim in life and that is to survive. They eat everything from the most poisonous plants to fruit to meat and bones of animals lying dead in the veld to dry hide and even one another. Fortunately too they only occur in the most severe droughts and when developing into the pest, which I describe, there is little to nothing for even grasshoppers to feed. Seeing the specie for what it is it made me realise that life somewhere after them made a choice to form complexity and variety or remain as they have and form a universal gene where the original mother is still present in all her offspring even after so many billions(?) of years having the opportunity to progress from where they were. They made the choice to remain the same where as the line that man developed by the original parents may have made the choice to evolve through complexity.

With all the explaining I do not wish to prove that man had a nibble or never had a nibble from that specific tree but I only wish to indicate that there are a variety of interpretations and clues around and when sanctioned they may deliver a vastness of possibilities. There is one other tree of distinction mentioned and also a mention of some eating by the female at first and later on by the male. This tree also was named and it was the tree of good and evil but according to the Afrikaans Bible the mentioning of the name says specifically and I quote: *Boom van kennis, kennis van goed en kennis van kwaad*". Directly translated it reads as follows "tree of knowledge, knowledge about good and knowledge about evil" and that is where my argument starts in my attempt to indicate the possibility of the sixth dimension belonging exclusively to man.

In the detailed analyses the specifics concentrate on the "knowledge" and then distinguishing between "knowledge of a good nature" and "knowledge of a bad nature". Please note there are three mentions of knowledge one only about knowledge then about the knowledge of the good and thirdly about the knowledge of the bad, but most important separating the three by distinction.

After clearing this part we may return to the animal world of some being wise and not very wise. Animals by nature and by genes acquire a base for knowledge to carry the specie through dangers and more important even, to the survival of the future of the surviving gene pool. Surviving as far as the animals go is the good and the evil and there, at that point all other definitions stop. All intelligence the specie holds and all the intelligence the specie acquired contains the one underlining element being individual and specie surviving.

In saying that I do certainly not say that rules amongst members of a group of animals is non-existent. There are certain criteria the individuals have to meet to establish rank in the tribe. At the same time such rules do not centre around emotions of ethics but they are practical well placed and directed to ensuring stability and it seems the higher evolved the specie became the more sophistication there are amongst rules constraining some members to the advantage of others.

The Matriarch in the Elephant herd is Boss and that is in capital letters! No bull will dare to push members and lesser infants around and she takes much less shit from young elephant males than the females. She is the rule and the law and every one abide by that. Should a male wish to afflict his attentions on some elephant cow the Matriarch will condition the visit to her satisfaction or the visiting male may even pay with his life.

I say this as a result from knowledge I acquired as some game-farming friends of mine has elephants in captivity. Crossing electric fencing is hardly an issue for the matriarch because she takes a teenage male place the young male (and it is always a male) between her and the fence and let him walk unsuspectingly alongside her as she deliberately holds him in a position where he walks between her and the fence with the current. Then at a moment she decides on she thrushes the young male allowing him to plough through the fence and take the electric current shocks while he goes on his way braking the fence altogether. After that the fence is open and clear for the rest of the herd to cross. She will never act in this manner towards young virgins but the young males get the stick every so often. With the advances the African elephant show would the African elephant be a good pet. I would hate to find out because they may set the rules and not me.

I know for a fact that a Nile crocodile does not make a very obedient housebroken pet. Should any one have an idea to keep one in his swimming pool be warned the pet would not distinguish very well between his owner and his next meal. And his love for children might be somewhat different to that which a good pet should have as is the case with dogs. With dogs man had thousands of years in breading good pets but it is unlikely that the first relations were as timid between master and dog as that we grew accustomed to.

1. Through many generations of exclusive inclusive breading did we finally manage dogs to have become what we wish them to be. In this lies another fact to analyse. Many different breeds make many different dogs and the one race has characteristics setting that race apart from other races but in the race itself the variety of characteristics find more prominence in the race than in individualism. Characteristics of dogs connect more to the type of dog than to the individual dog and therefore some races have inborn hunting skills where others may have guarding skills. It is the breed that brings the selection and not the individualism in personalised characteristics. Therefore it cannot be said that a dog has a conscience but it is better said that a dog has a better breading line.

2. Some evidence suggests that when Cro-Magnon – man arrived agriculture replaced hunting as the feeding method and we are confident man exclusively kept that dog for it's hunting and sharing abilities. With the arriving of agriculture man then extended his space-time manipulation not only beyond his physical abilities in hunting but also his physical strength in working with tools. This must be the biggest leap of all even much bigger than the leap of the electronic age but such comparisons are extremely difficult to make.

What would be man's drive to not only manipulate his personal surrounding but also manipulate surroundings of other forms of life to their benefit but moreover to his benefit. What would give man such judgement as to select species beyond him and feed them to eventfully find more benefit from their feeding than they did benefit from their feeding. Genes it cannot be . The orang-utan has 98 . 2 % of the genes man has and the gorilla has 97.8 % of the same genes man has. The Gorilla still lives in woods and is destined to disappear while the orang-utan lives in trees and holds no better future prospects. Genes would at least give the species having such close relation in the gene pool with man an idea to follow the trend set by man and copy some of the abilities. Genes it cannot be and that just about excludes the last cosmic or natural physical explanation from the list of possibilities.

There seems to be a massive gap between what man became and what ape became. Science makes a great singsong about chimpanzees with the ability to use tools for their benefit but man has surpassed that so long ago science have no tracking record about the time and the way that came about. It seems as if man was not, then man was with agriculture and all other providing the manipulation of the other species under the control of man and by increasing all benefiting that the animal enjoy man could bring benefiting all around to benefit man.

One day man was ape and the next day man became super-specie-of-the-world, the world champions in space-time manipulation or in other words of controlling life. Not only the life of man but also life of others to some mutual benefit slanting heavily in the favour of man. Still the benefit of the other species holds so much that only species that benefit man started to dominate world population with man.

Of course as usual and as with most of Newtonian scientists' findings, I question the accuracy of the gene pool percentages strongly and I am of an opinion that such percentages are in use for political issues more than scientific proof. I prefer leaving it at that. I am a very small fish in a very large pond sharing the pond with very powerful other fish that can destroy a small fish like me with one gulp.

Going into the development journey as man followed the trail one has to look at not what man achieved by own ability but with own measure in manipulating others in life. When man started with chips and flint it was progress but it was also very limited. Only when man acquired the muscles of more powerful animals and took their ability in measure with mans manipulative power did success arrive at a level that brought progress in leaps and bounds.

It is not mans hands or legs that brought man's domination and control but man's brain that brought response from other life to benefit man and find benefit in shared life styles where mutuality brought about safety and mutual prosperity. It is surviving more than anything else that means good or bad to animals and most of all surviving of the species and in that the animal only find man and man's company good because man holds its safety. When a lion brings down it's pray the rest of the flock will start grazing immediately without showing even slight remorse to the victim and for the loss the close relatives are faced with. By the death of one the rest find safety and to animals that is good. To the rest it is about surviving and if one pays the price, little concern goes to who paid the price. That is what annoyed me about our all-famous-pop-star-wife. She truly go beyond what nature puts down as rules applying her liking (and selling her book with cheap bluff) while other sheep walking on two legs like only humans ought to cheer her stupidity as if the stupidly were they're own. They have not even got that much brains to acquire that much stupidity but has to borrow to get their tally that High. The biggest annoyance is that with such stupidity those can vote and choose my future because of democracy. They prove almost all ways not to have the thinking power of a mouse but they have the rite to choose my future and I have no say in the matter but to follow what such morons may wish upon us.

With the novel idea (novel as it is only man that uses mutuality single-minded and still provide beneficial good from all angles) of widening the use of abilities provided to different developed species, man gained extensively in progress and comfort. After all it is much lighter work riding a horse than walking all the way.

But gains in comfort goes both ways as the horse find protection against predator attacks while finding good nourishment in winter and the best hay to feed. Such a diet provides the horse with the strength to carry the rider and enjoy own comfort with the fact of much reduced fear and anxiety. By having male and female and promoting mating the good in the life of the horse becomes better. Did man rob horse from freedom? Did man take what was not his to take? Should man get some conscience attack leaving him with sleepless nights about his cruelty in robbing the horse of a natural life of freedom? Well the humanists will tell you with teary eyes and running noses that the horse should have its freedom as all are born to be free. But this emotional outcry comes with the comfort they, the humanist enjoy of secured sleeping a good all year round food supply and breading safety. I have not seen one humanist go running into the mountains never to return to civilisation, to enjoy the freedom they should enjoy as much as they wish that upon the animals in captivity. To the horse, after the initial fear of the subduing and the ultimate realising that the subduing is not life threatening as he can live with that, he finds comfort and even enjoyment in that. I see on a daily bases how horses get jealous when the owner takes one to ride when the rest wants to go first. They come and nestle with a desire to connect and in jealousy push each other away to be the one receiving the owner's attention. It is a bigger issue of the conscious to decide the likes of others in what they like and what you think they may like. No one of sane mind will hurt an animal in your care and when slaughtering that we do in the most humane way as described by law. No one cuts of a chunk of steak while the animal is alive. We humans have civil norms and values and by using our brains we can live and let live with more dignity going around sparing the animals huge cruelty than what the animals would have come to face if they're fate still was in freedom and being hunted down by wolves and hyenas Such is the difference between those having bleeding hearts and brainless skulls and others that can think. Now we arrive at an interesting question as how do we think and reason. I am sure all humanists will have as much to say about my way of being correct only as they will find many arguments as proof of the fact that by dislodging my logic they can prove me being beyond the norm of classifiably insane. There is ever a clear definition about rite or wrong and all principles we find appealing or appalling is within the brain

According to an article I read the brain holds more connecting lines than does the universe and I may even accept such a statement on the grounds that life has much more complexity than does the universe. After all life takes the dimensional barrier as far as the universe does and then beyond where the universe stops. This does not make the universe simple because I cannot see how any person may ever come to understand the flow of light as the light uses both the straight line of singularity, the half circle presented as the Roche limit in singularity and the Titius Bode triangle making light representative of every aspect which connects space-time away from singularity with singularity as light where Π meets r to become the value of C. But in the brain this is only one function as electricity holds an equivalent of light forming electricity as the messenger to whatever energy is above life and then in the human capacity above even what forms the barrier to life. The arm is not human life because a human can loose the arm send still be alive. Therefore what ever is in control of the arm is in control of life, which puts what we find as life at a higher dimension than that of life. You may argue that in case of animals such thought also control life because a dog may lose his legs but not his life. But even as complex as that may become there are relevancies between life and the physical because where the physical uses pain as a warning system the mind uses fear as a warning system.

By following such a line of argument one can freely deduct that an insect as our corncrake, which we discussed earlier on, is representative of life as the same life we find in the arms or legs of our body and the life we control but is not truly part of the energy "me". Clear to all it must be that life we find in mammals are advanced above and beyond the development the insect arrive at. If that is the case then I may claim that human life has more developed than what other mammals did because with my manipulation of space-time I may manipulate other mammals to harvest some of their manipulative abilities in benefit of our mutual relation inclining more to my benefit. The cow does not seem to mind when I milk her but can any person imagine experiencing a milking session involving a crocodile? Well, fortunately crocodiles have no milk but if they had I would never volunteer for the honours of being the first to train a crocodile how to behave in a steady manner when in a dairy session. You may have or may not have noticed but I am telling you that they have a sharp side and they have a blunt side and the sharp end holds rows of teeth they surly know how to use. Even the blunt side hits like any whip never can and I am sure a fully groan specimen may kill with that tail. Going down the order of evolution we come to bacteria and viruses, some of the lowest forms of life. Do bacteria and viruses count as animal and if not then surely they count for life because life they are. In that we find the equivalent of bacteria in higher developed species as we find that the insect may have the developed life mammals use in their bi-products included sustaining their superior life development. We can see evolution by applying a relevancy of devolution to siphon and separate life from life. I am trying to indicate that life becomes a compliment where the lesser developed formed a mutuality and aided the supreme form of developed who is controlling the master brain in that form of life wherever the master brain may be attached.

Life is above and beyond the cosmos and surely even the most ardent supporter of atheism must grant me that much. By that grant must the atheist then add the fact that life cannot only fall onto a category of to be or not to be but there is a range in life forming a line of development and superiority. Life is more than life but has status of being lesser or more and that is the point I wish to address after all the talk.
Within one body a range of life values combine in making whatever accomplishments the life form accumulated by extensive development that range in development. It is appreciable in concluding from a range of facts I mentioned but mostly from human common sense we all know that on top of the range being the model best manufactured and with all accessories all other models also having life envies and fear is man. What will make man that special?

In 1905 a case was reported for the first time of a woman that had a hand, which attacked her every night by trying to strangle her. In the manner the hand acted it was clearly out of her control as it was clearly out of control of whatever controls the brain have over body functions.

She would wake at night and feel someone strangling her but the person strangling her was something she ought to have under her control. Imagine waking one night feeling someone squeeze all life out of you with a murderess motive. Even the thought of that will make most people get up and bolt their doors and windows just in case. It must be awful having a murderer wake you with such a horrific intention. Go one step further and think that person may be one of your house members you trust with your life. A thought of that becomes rather preposterous! Then for the ultimate in revulsion; think of the chances something acting in such a horrendous manner is something you know with every grain of your body as that thing acting is your body. There is nothing worse to be scared of than being scared of "you". How do you fight such an act. You cannot hide and you cannot go without sleep and as you go to sleep you know that there is some part of you yourself hat is after your life. If this is not enough to drive any person into hysteria I do not want to know of anything worse.

This flabbergasted the doters and no one seemed to make any sense of such phenomenon. I suppose if such an incident had occurred before it would have been denounced as an act of a demon of some kind but fortunately for medicine the art of healing had abandoned forces of nature as a scientific accepted fact unlike the likes of physics still clinging on to such madness. If my memory serves me correctly this case was in Germany. Please remember unlike our distinct academics I have no extended library to find all kinds of information but have to rely on a failing memory being destroyed by my diabetes.

Then in France later on another case became known about a woman that had a hand also out of control where in this case that hand tried to forcibly turn the steering of the motor car she was driving to force an accident of a serious nature. This manipulation was seemingly as much out of control as was the previous recorded case. The common factor about the two cases was that in each case a person had an arm that was intent on destroying that person without the person aiming to do so in free will.

Later on in America two neurosurgeons planned an operation procedure where by they aimed to relieve patients having chronic and continuous convulsion attacks caused by epilepsies in the brain. These cases were dire and with the operation as a last resort all the serious after effects became a secondary factor to the superior motivation of saving life and improving the demented quality of life by the patients .

The operations involved only the utter most serious cases that left no other option for improvement. It was this or death and not choosing death the patients chose the intended operation procedure instead, but still it was extremely serious and dire options in the choosing.

They reasoned that the epilepsy was a result of the brain having vibration and with the vibration stimulating other vibrations through the brain in some cases the one caused the next vibration and it was more a reflex of the first causing the next as the symptoms was going on a prolonged non stop convulsion. To stop such reflex by the brain tissue they held the argument that when cutting the cortex the two lobes attached will not have the reflex and thus the continues convulsions will loose the continuous effect.

By separating the lobes the nerve attack coming about in the one side will not transfer to the other side thus it would not cure all elliptic attacks but the prolonging effect will be reduced. One vibrating lobe wills then being separated from the other part not cause the response in the next lobe and it was diagnosed that it was more a response to the reflex allowing a reflex to the respond and this brought about a never-ending cycle. The idea was that it would result in reducing the severity of such grand mal epilepsy

According to American law the doctors first had to show a high degree of success by operating on rats in order to prove that the consequences of such an operation is in acceptable levels before starting such a procedure on humans. To obtain the rite by law for the granting of the operation many rats underwent the procedure and the procedure then were extended to many other species.

Every aspect of recovery and side affects must be documented to an exact accuracy with no exception to the rule in the slightest.

The behaviour of the animals before and after and the general physical data then goes to excessive detailed scrutiny by the finest the medical profession has to offer in America. Accuracy in the process of accumulating data and other relevant information is beyond question especially in the country with the highest standard in medical care.

The after affects the procedure had on animals were indicating no serious side affects of any reason for concern. Many species went through the procedure eliminating defects if whatever possibilities there may be.

When monkeys went through the operation procedure our primate cousins had no side affects in any way. There was perfect hand eye co-ordination and the nerve system had no complications with the motorized operating functions in any way. This confirmed the surmise the medical profession then at the time had that this third lob was just fibre with no function of distinction. The fibre was in position to stabilise the two lobes and had no connection with the lobe in a functional manner at all. All the indicators brought about such positive results that the American government granted the licence for the first experimental operation conducted in a human in absolute confidence the procedure went about and with a very good outcome. But shock was looming to all medical experts.

In every case the patients had one common disability. It showed a horrible disadvantage no one expected in the least.

All patients showed the science of a phenomenon later named after a movie Peter Sellers made famous. It was named the doctor Lovejoy syndrome because on of his characters in the movie was an eccentric half mad all crazy German general that had one arm always trying to strangle him. This was meant to be funny in the film but the patients suffering from the reaction of the aftermath are not so inclined to the humour. They all had one arm that went out of control and the arm showed serious signs of having a mind of its own by doing the most annoying things the patients had obviously no control over. The one arm had life apart from the person free will doing things that would embarrass or even threaten the so-to-be owner of the limb.

Well I am no brain- surgeon although I am inclined mostly to form an opinion of my own that may not always stroke with informed opinions by professionals. The test operations were conducted on a variety of animals including monkeys, the so-they-say close relative of man. Well as close as they can get but I am of the opinion there are other species being still closer to man, but that is somewhat off the point. From all the facts I mentioned the past pages I drew a conclusion of my own.

I showed that man has a higher evolved form of life than other animals.

In the Bible it reads that man was made the last but far from the least with more superior qualities than all life combined because man is all life combined and then added more than the fare share. Any one thinking of our Creator as a magician is mad and that I state without excluding even the Pope or preacher of whatever denomination. The Creator is a building architect applying mathematics and physics we can never come to appreciate. If there were any one that has an opinion that God spoke a word and magic was the word that person would have another opinion when thinking with some clarity about Creation as a whole. The Creator is Creating and by creating there is a building process involved. Every person starts an individual process of building a human just after birth.

Looking at tribes living in regions far away from civilisation as one can still find in the Amazon River we find those individuals being adults by body still play games we find our children play with much amusement and childish enjoyment.

It is far from incomprehensible to make some sort of comparison between our children at play and those adults at play and the similarities are astonishing. Racist remark it may be but the grown ups are more child than the western child is child. They defiantly are backward in mind and mentality. The most logic deduction is that the child in us represents our development phases through a long journey. Humans at birth are animals. Babies can make noises to convey their needs and nothing more. As the little life grows it develop not only by growing but also by culture and what the parents put into the culture. It is more than likely that the developing pattern children follow is the same developing line humans evolved through as the generations brought more insight and better understanding to the following generation. It is only of late that there is some pattern of devolution taking place especially in the western world but that trend is set by a culture of greed. The parents are chasing a good life and instead of placing morals they push money in the hands of their children to get the children out of the way so that the parent will lead the life they choose and rid them of the burden of children while conveniently blame teachers for the children having unacceptable behaviour as much as they pay psychologists good money to correct they're mistakes. It is moreover a fact that parents do not actually pay the professionals to fix but pay the professionals to rid parents of responsibility, guilt and of course the children. The price society will pay for such luxury is far more expensive than affordable.

I so many times wished I could have my life over again. There is one study I shall conduct and that is follow the pattern the child indicates how the human developing process came to pass and draw the parallels from that to man's evolutionary path. It truly must be a study worth one life. To know man that well must be the ultimate there is to know and afterwards death can bring no regrets for wiser no one can ever be.

I think we should now return to what the Bible refer to as the tree of Knowledge and reflect once again on that verse. The quote was and I quote in firstly Afrikaans: Boom van kennis, kennis van goed en kennis van kwaad". Directly translated it reads as follows "tree of knowledge, knowledge about good and knowledge about evil" and that is where my argument starts in my attempt to indicate the possibility of the sixth dimension

From the stage of toddler I found eve dice of three characters in all persons. There could be more but not less. Let us call the three persons three entities of the good the bad and the ugly, but not in such a specific as being a saint a pleasant and a demon. The entities are rather more under cover that that straight foreword by definition. In every person's brain somewhere there are the trinity of entities ruling our lives.

The trinity are one person and not individual persons but the same although very integrated they are also very separated. It is not a question of schizophrenia with multiple personalities because I do not believe in that. I think that was made up to convince who ever needed convincing about something to do with nothing and has no base or then has the same base as physics hold gravity responsible for a variety of facts they otherwise have to admit they no nothing about.

The three belongs to one person and in fact is the same person. In the Afrikaans book I named them Ek, my and myself translated as I, me and myself. To make matters more interesting and subdue confusion somewhat I wish to keep the Afrikaans as that would make explaining slightly less complicated. Ek is I. My is me and myself is will you believe it myself. The "My" one pronounce just as you pronounce the month of May in English and that would make the pronunciation of myself as you would say May-self in English with the pronouncing of self in English and Afrikaans exactly alike.

Ek my and myself are the same and there are no distinguishing between the characters but at the same time they are as far apart as three that never met before. Every of the entities belonging to the same body, the same brain one has a character as unique and as far apart as another being on different sides of the universe.

Every personality has different likes and different dislikes and feels a different purpose in life as much as to life Once again I wish to press home the fact that this is (to my view) as normal as breathing and has nothing to do with the mental instability known as schizophrenia. Although only one personality claims occupation of the body at any given time, all three take responsibility for action all the time because all three are the compliment of one.

Any one such personality may claim occupation of the body at any single moment and normally do not relent occupation easily but of the three one is always in charge and the other two take position when the main personality loses concentration or relaxes guard of the situation. Every personally has own motives being apart as far as the north may be from the south. One may even think them in classes of being the person's personal god and personal devil but such a thought may place boundaries that are unfair.

I would rather describe them as one is the charger being scared not even of the devil himself and the other will be the cautious the guard, the one always on the lookout for trouble coming. They form the one that is in charge and the one in charge cannot be excused because it was a totally foreign entity pushing in charge in the direction it never wished to go but was blackmailed in doing that wrong! The one in charge takes responsibility to the full for each dead the body did and every wrong committed. In the end they are only identifiable but not that clearly deniable and they always appose each other in complimenting one another.

In the American cases where the neurosurgeons performed the operation in cutting the cortex to split the brain lobes of the operated patients a condition became a situation where the persons found the uncontrolled motorised motions of limbs under their control supposedly, but not under their control at all. Then for the first time the phenomenon became a syndrome with a name. It became "the alien hand syndrome" but also find referring by the use of the Doter Lovejoy syndrome named after the movie. There are cases known on record to result from a variety of brain damages and severe apoplexy. I can confirm as a witness that in the cases of my biker friends their faces changed with mood swing. Not bone structure or complex feature but the facial expression changed the way the muscle form the face. In cases they looked somewhat alien, but my referring to as normal is not about such extremes.

The alien hand syndrome observes cases where the one and the patient refer to as the naughty hand and the clever hand. The naughty hand seems as if it had a life and mind of its own sometimes acting to embarrass but some times even to endanger. The main connecting issue between these cases is the hands control not being within the owners authority and that the hands will obey as nothing ever change but then on occasion from the blue it will act on own impulse and with a motive clearly never matching the owners intension.

I wish to underline with no exclusion to whatever intension that the alien hand syndrome does not prove the trinity that I refer to and the alien hand syndrome underlines two facets where as my view state clearly three identities. The reason for this mismatch holds a most intense connection to my personal religion, which I never share, with any person out side members of my immediate family. The alien hand syndrome only confirms (to me personally) a connection of some sorts in some way.

Many years ago after I had to overcome some personal problems at the time through which I was admitted on occasions to an institution. I spent time in nerve clinics where I had to recover from some brain disorder called endogenetic depression. It is a condition that the patient will get severe attacks of depression and the cause of this is totally inherited by nature. The disease comes from a bad gene carried from parent to child and as much as my father suffered from it so will my children and is not very scares or very serious. I would say it is as serious as you make it to be. My regrets about suffering from this condition is minimal because of the extent of learning I had the opportunity to come by that otherwise would never have come my way. But then from what I saw in the times I was admitted to the clinics I mentioned, my (unprofessional) conclusion is that in just about all cases the prognoses depend almost entirely on the patients willingness to recover and how serious the patients are about recovering as the recovering is al about the relevancy struck between obsessions and true problems. In that there rests a balance between fighting for the sympathy they wish to evoke in others and fighting a battle to achieve oppugn from the problem. In hindsight and after all the bad is forgotten it was worth wile because the learning of the human mind and the way others think and feel was enriching beyond my suffering. At the time the suffering was almost overwhelming but to escape ones own problems one can listen to others and by learning from them one get perception about their state of mind and your own state of mind. It was in this period I came to the conclusion about the trinity within us all. By increasing my personal learning curve I decreased my personal discomfort.

Inside all of us lurks to personalities above and beyond Ek and they go by the names of My and Myself. Most of the times they behave well but sometimes they can bee opprobrious without my consent in the matter. One of them and I leave the choice up to you are truly obnoxious and spiteful more to yourself than to others. The other one is normally timid and can be classified generally as your conscience. Between them you come and take the control because you are the boss. Being the boss and demanding control you are the balance as much as you take blame and shame when not being the boss and not in control where matters go out of control. You carry the consequences always as you should because whenever what ever goes wrong others will pass the blame on to you. And so they should because you are responsible even in such times that you are not responsible.

Ek finds himself in the middle of My and Myself and as Ek is in the middle Ek get advise from My and Myself. In accepting or rejecting comes regrets and jubilation but ultimately the final choice is with Ek. The characters are strong at times and are weak at times limbering and dominating whenever opportunity presents. In the Afrikaans I identified the character by names but found it somewhat complicating the issue in the English and so I did not name to identify by character.

Ek takes full responsibility for all actions for Ek is My and is Myself by only being Ek. We all have this in us and some find ways to fight it better and some of us are in a desperate fight for sanity and survival. The issue is not the guilt because every one is guilt ridden as we all have to cope with this behaviour in some degree. The stronger any one denies this the harder the one is in a fight for self-protection. Cases are sometimes most serious and other times just under the skin but it is there in every one. If you are human you are fighting. Being the one finally carrying the burden then becomes self.

It could be where he father feels he as person never accomplished anything and with that self detesting he puts all his blame, guilt and rejection for what he is into the passion he feels for his child. The passion he feels is carrying the burden of hate he has for his image and that drives him to expect from his child what he never could achieve. The child has to be many times better performing with many times more positive results, be a champion, be a scholastic genius, become the school president and make a mark as a pillar of the community although the child is only a child. The child has to outperform all others on all terrain in a fashion fitting the image of a champion, which the father never was. It could be as serious as I indicate or it could only be in little suggestions made to better his child.

These same feeling can bring about the very opposite where the father tries at every chance he gets to run his child into the mud because the father fears that the child will push him out of his role he has to fill but lack all abilities to fill the role. By destroying the child he is protecting he child because he is securing his position as the father figure. Knowing well the chid carries his genes and believing the genes the child is carrying is not worth much, he protects the child by showing the child what he (the child) is and allowing the child the realisation before the child will one day find out for himself in the cruel world. There is no good as much as there is no bad, and in the same breath everything is wrong.

The same behaviour may come from the father in fortune, the one with success, the pillar of the community. He is the top judge, the success fill attorney the town's top businessman the city mayor the admiration of others. He hates his child for that child will one day inherit all the goodness he now has and that makes him sick. He might feel the child will never become the doctor he now is but through his reputation that he worked so hard to achieve will become even greater than he now is. That makes him to push the child to live up to his personal greatness or destroys the child to show the child how much the child should be great fill for the admirable fortune the child has to have a father such as he.

It does not have to be about someone you love. I shall be very frank about my case and millions will recognise their fight.. My personal struggle involves money for one. I have not the slightest idea how to administrate my money affairs and sometimes I know I have this inward hate towards money. Whenever I have money I allow people to sucker me with a sob and crying story about their hard life and the bleeding heart dashes through, the knight within me with the glittering armour takes full control and helps me to give away sometimes even thousands and tens of thousands, knowing very well notwithstanding all the promises of repayment at the time, I shall never see the money again. It is not I being the bleeding heart and then yea it is I the bleeding heart but that is one character. I the bleeding heart am on the background where I the hating bastard am rite on the dot standing on all fours and then some shouting and urging me to help the others, or buy what ever. Forever it is ensuring me that there is thousands more coming my way in any way so what the hell, let some goodness flow. This will always come where for some reason some money fountain runs dry just afterwards and I drop my family into financial surviving periods allowing them, the ones I love most to suffer the hardships. Not once did this occur and was not the prelude to personal hard times! I am a middle-aged man. I know these characters. I recognise the precise feeling accompanying each one. I have scrutinised and analysed they're being part of me decades ago as I and the bastards still catch me again and again, sucker punching me over and over.

The first time I went gambling I recognised the one being very aggressive and on the forefront just under the skin filling me with anxious excitement and then and there I realised gambling was not meant for me. In that sense I have beaten him hands down by not starting the habit. Old positive me is about cars speed and going crazy in my mind and I know it is the strongest one because that one am the biggest I. A middle-aged man tearing down some street on a massive motorbike showing some youngster what the bike can do when there is someone on it that knows his onions are grossly irresponsible. How childish can you get, how in mature can one be. That is the I and that is the very me and that is Ek and I have as much control over that as a drunkard has control over his drinking. Saying that is also admitting I do not wish to fight him as I do not wish to beat him. I find his company very pleasant stimulating and destructive and when he comes to the foreground all other characters harmonise giving me all the different feeling each one of them should supply. I am then self-destroying giving the negative character his day, I am childish giving the neutral character his day while the positive one and I am the same person to let all power loos.

My characters bring me in conflict whenever I bay or sell or demand a price for my services rendered. I would bay the biggest shit at the highest price in the belief I am helping the poor slob. When I sell I get all guilt ridden when trying to make a profit because I get this idea I am cheating the poor fellow by insisting on the price I am aiming for. In all cases I feel so guilty to ask any person money that actually belongs to me I get a nerve attack or a running tummy. It is not out of fear because if any person gets violent or aggressive I know how to defend myself. You have to know that if you are a devoted biker because they always get drunk and strong at the same time and fortunately I was in motor racing during the age my mates learned to drink. But with racing cars always braking down and crashing there is never money or time to drink so I never got around to start the habit and by the time I stopped racing every one accepted me as a teetotaller so I never got around to develop the habit.. But asking for money or insisting on a fair price is more than I can achieve.

I shared with you this Ek, My and Myself part of the personal me-story so that you, every you will know what I mean because the objects may change but the objectivity and feeling and the motives never change.

One example how the process works will be when a mother that hates her child for birth pains, unwanted pregnancy, feeling unfulfilled through a bad self image or carrying on where her parents left off with them treating her as a child in the very same manner she treats her child. The negative character prompts the mother to reject the child as she detest the child and loathe the child. The Positive character saddles her with a tremendous guilt punishing her as anxiety in blame riddles her.

In self-punishment aiming to destroy her because after all that is the purpose of the negative character the positive character punishes her as she fills her with self-loathe while neutral character tells her about the wickedness in her. She gets reminded at every opportunity about her love she must have for her child and her duties as a good mother to love her child. It is her responsibility to love and protect and because she strives to be the good in her the punishment is severe.

If she gave in to the negative character and start enjoying the hate and blame she feels towards the child the positive character let loose the pre-historic maternal instincts with a flow of torturous guilt hidden behind such strong emotions she deflects the hate onto her self.

Then comes the neutral character reminding her that no one should ever know about her hate towards the child because as she detest herself so would the world detest her if ever someone became wise to her feelings and she will be driven from society with hammer and tongs. With the conflicting hate polarised and swinging between her and the child she knows that all of human kind will see her for what she is and with the hate she feels towards herself the negative character takes that hate and turn it into fear for others finding out about the truth. In realising that no one may ever know about her true feelings toward the child and know about the loathing she feels about herself she takes all precaution to hide it from the world.

In this mind game of rocking emotions the positive character supply her with advice in how to take charge of the situation never to the benefit of the child but to her benefit in protecting herself from the outside world. The advice will never have any concern about the child because she carries that burden by herself. The positive character continuously reprimands her of her evilness while the negative character fills her with hate running between the child and her feelings about herself. The neutral character reminds her constantly not to allow any one to find out about her feelings for her child and demands protection of the outside world finding out about her and her child. The neutral character pushes the fear to match the severity of the onslaught by the other characters in order to maintain equilibrium and equilibrium means almost insanity

As the insanity at times almost become intolerable she reflects the blame for her situation onto the child being there and making her life hell. This the negative character grabs by prompting her to punish the child as severe as she can and through this stop the child torturing her. The cycle leads to the next cycle and in this sanity guilt love and hate becomes one flowing emotion of disturbance. With this conflict within the mother the child's developing personality receives knocks the child cannot stand and less understand. The child starts behaving rebellious and unacceptable to the mother's neutral side and in the eyes of the community. This chance the positive character grabs and being positive only as far as the mother's well being goes the positive character advises the mother to leave the child and let be. This will show the world how much she loves her child by refusing to even punish the child when the need arrives.

In this advise the negative character joins by advising her to let the child become out of control establishing the fact that should her hate towards the child ever leak out, the world will not blame her for every one in the world that has contact with the child will hate the child in any case. When every one despises the child the negative character swings into action by filling the mother with more hatred towards the child and when opportunity comes and they find themselves alone the hate comes to the foreground and then she punishes the child with most cruelty. This can be as part of actual criminal prosecutable child brutality or it could be most cunning and devilish in conspiring but the brutality is all the same. With the presence of such severe child brutality all others in the community turn a blind eye not to get involved where each outside person will find some excuse not to become involved. As every one has a struggle of their own they too are in self-protecting not feeling the urge to come into the open and defend the child. Being in the open will unveil their personal fight for survival and they then will become the target of the community. All this is in the very distant back of our minds never in front where we can kill it but present in the way to be us and not to be us.

All three personalities agree on one thing and that is that the world must with all its people have a hate in the child as much as the mother. But as every one find the child unacceptable and revolting she can feel better about her feelings because now she is part of the crowd.

Being part of the crowd will bring sympathy from others with their understanding about the hell and the torment this child inflicts on her every day. Such a feeling soothes the aguish of the wrong she feels she is committing as much as the wrong the child is committing to her. By allowing the child adverse behaviour and defending the punishment of such behaviour the child will become more unacceptable and the situation heads directly for the disaster she hopes to accomplish. She directs the child's personality in that direction while she feels all the torment others see her go through. She allows the child to go to nightclubs doing drugs and commit self-destruction while the mother is merely a spectator because after all that is the child she hates.

When the child is out at four in the morning she can feel good about not having the pest around. She can feel good about her hating the child. She can feel good about all the sympathy she receives about having such an unruly child. She can live the life she claimed, hating the child, feeling sorry for herself and good about others understanding what she is going through. Should the father, a teacher, a policeman or any other figure try to stop the madness and bring order to the child she will attack that person with all the hatred she feels towards herself and towards he child. She will destroy the prevention of her self destruction because any body trying to discipline the child is fighting her aim to destroy the chid and after all then the world will see how she can fight for her child's protection by almost putting her life on the line.

This becomes the Sudan affair where the bleeding heart buys guilt relief and be god while the philanthropist pushes guilt as hard he can and be god to collect money on behalf of the Hoggenheimers that then can be god with such wealth distributing it to the Mammonites who can be god by baying from themselves as much as selling to themselves with unscrupulous profits making him god and allowing the Mammonists to be a slave driver and being god to the slaves. It is this sickness of society no one cares to see because every one gets what they want, even the luckless get what they want with the minor condition that when the luckless suffer most that becomes the region where most profits are for every one in the chain of gods. So the luckless must be in crises starving as they are dying to gain most profit for every one. The profit has little to do with money but with being god. Any attempt to stop the situation will never be tolerated by any party and therefore my remark that every one will press for my castration because of my suggestion to rectify and bring a solution.

The mother will fight any and all positive solutions with tooth and claw and no one should dare to lay a finger on the child because she know how successful she is in destroying the child and it is so easy to shout child abuse when someone wishes to correct the ways of the child to the benefit of the child in the interest of the child. Her devoting love will protect the child from such brutality as what a good hiding on the backside will bring if the hiding is done in love and the child knows it was on behalf of care. But that will stand in contrast to the mother's brutality and punishment and the child may recognise the difference and wise up to the difference.

The social worker will never accept such brutality after all it may cure the little brat and put our social worker out of a job. The lawyers and judges are on such a big job creation drive by minimising penalties and getting criminals back on the street for the next cycle in crime they will fight any interference that may reduce the crime and decrease their chances of money making. With so many to loose so much on the one side and only the child to gain on the other side all brutality in the name of positive punishment will become child molesting and will never be tolerated by those with influence in society.

The mothers behaviour becomes a reflection on the disease within society and it is in everybody's interest but the child's not to admit to any knowledge about the foundation behind the scenario, after all it is only a child going down the tube and to top it all it is a child no one cares for. Is there anyone out there that can see the parallels running here between the animals not caring for the unprotected infant crying in desperation for a mother while every one ells in the species cannot be bothered? May I now comment on the fact that we are going the way of the animal and devolution of our species is in progress?

Weather you care to admit it or not but greed and money is destroying man to the fullest while man is enjoying the destruction with all its lust. The drive in society after W.W.2 became progressively to feed the children to the hyenas of society, which are the crime bosses, the prostitution rings, the drug pushers because after all, they bank the money at the Hoggenheimers to the convenience of the Mammonites and Mammonists. The ones that are caught in police action are the ones not part of the official system and they become the offers the politicians demand in protection to show the public the system is doing what it can but unfortunately it can only do that much and if it is not good enough it is because we all are human. Hundreds of thousands of children disappear through out the world and no special task force has ever been set into action to get behind the problem This problem has no boundaries in as much as it is going on in every country there is world wide. I have an awful, awful feeling and please consider the next remark as a thought with no substance but that internationally oil is bought with children as payment because no politician through out the world shows much concern. But let three banks get robbed in one day, then a special task force comes into action and gets the culprits with extreme prejudice. This is the symptom while the mother's behaviour is the condition.

The positive character advises the mother to defend her child against disciplinary measure, the neutral character reminds her of the image problem and demands protection while the negative character sees the child slip into the ditch and everyone is happy. Every aspect is in line with almost one aim and that is the destroying of the child. What ever may bring positive results everyone shout down by making the connection where the punishment links directly to what may be extremely negative because from the onset it seems cruel and negative. Giving the child the spanking of his or her life driving the fear of god into the young person will have extreme negativity but that must stand in complete contrast to the love the child then must receive before and after the spanking. The child must know with one hundred percent certainty that the parent is and will always conduct the child's care with one aim and that is to ensure her or his well-being. But if the child knows that every aspect of the child care swings around the drive as far as getting the young person destroyed the yes, the child will find all connection to punishment intended on the destruction aspect but it will remind the child of the similarity and lack of contrast in the usual treatment because of the absence of the love and caring aspect is in harmony with destroying.

The whole aspect changes around when there is another person also punishing the child but with prejudice intended. When the punishment comes from despise and not from care the mother sits back and allow this to happen. If the father is sexually abusing the child the mother will suffer greatly all in silence all quiet not allowing outside intervention spoil the situation because then it is the father who is to blame. It is the father that is destroying the child and it is the father that is the devil. The other parent can then take responsibility and blame and the mother becomes the second blameless victim in the case where she does not participate in the abuse only because the father then plays the part. That is the only aspect that changes where as otherwise the scenario remains the same. All characters play their part as if she is doing the destroying because she is doing the destroying by helping to provide the perfect environment for the destroying and not allow any clue get outside the close knit intimate family circle. She takes a part in the abuse and takes as much enjoyment as if she was acting although the blame and the soothing shifts somewhat but all intensions still encircle the destruction of the child.

The conflict within the woman may drive her to protect the child she hates as much as the father does. Her neutral character tells her that now she is no longer to blame therefore what ever happens she can stand in the shadow of the male and he has to take the blame. But the negative character will not tolerate such idleness. The negative character wants the child's destruction but moreover the character wants her destruction. He advises her to action. When the father is at his most dangerous being overwhelm with cruelty she will jump in and save the child by physically protecting the child while knowing full well that she can do as little protection as the neighbours budgie can. This action will satisfy the positive character by her showing her unbound unlimited care and devotion as the epitome of true motherly love. With her actions she know she will unleash much more anger and the father will loose all control. The negative character finds stimulation in this action and supports more involvement to unleash more violence all the way. In such a rage she knows the father will then beat the daylights out of her before he turns onto the child with more rage than he had before. With such reaction all three characters are satisfied The positive finds a way where she can become good, the neutral character knows that society will condemn the father and the negative character will justify the cruelty as the correct way to go because the child the father and the mother is bent on destruction. Now the positive character can tell her she did what any good mother would have done, the neutral character knows the beating is the same as what she endured as child in any case and that did not kill her so this beating cannot be that bad while the negative character will enjoy the situation to its full as she and the child is being destroyed.

The balanced behaviour of a person with the trio not having violence to promote would run outside and call outside help from any source available at that moment. A woman beater and child abuser is always, always a coward and when real trouble arrives he will stop immediately. But if the trio is involved in violent provocation exemplifying hatred to the child, she would take charge in a different manner. Even if she does act in this manner she will not call the police or get the husband behind bars because she argues that the family will suffer with the father not providing at the time. No one receives any money while being locked up. The excuse she uses is that there will be will no one to provide for the needs of the family living expenses. The father's inability to provide is all but the truth as she wants the situation to continue because she is enjoying it as much as the father. Should she truly admit to the seriousness of the crime she would have the father locked up as if he died and never allow him close to any member of his family again. To her and her child the best will be if the father is dead because the father will destroy where ever he involves himself. The only sane thing to do would be to declare the father dead as far as the family concerns go because when he is released from prison she would have a new life to live that no longer will depend on him or his providing. She would recognise him for the monster he is and not see him as the senior partner in crime, which he truly is. Keeping this partner ship in place by using a lame excuse like ninety nine percent of woman finding themselves in such a situation uses would satisfy her three characters and once more the money matters more. Now the stage is set for the beginning of the next round because with the father as with her the creation of the next climax begins and develop until the next time hell brakes loos where the father then has even more hate and a lot more to prove and correct. His fury and outrage will cover his hate but also the negative character in him will demand revenge as compensation for lost pride. He will repeat his role and she will repeat her role and the child will have no role but suffer destruction and again and again the process goes on and on.

By throwing her body in between she knows no woman can stand against any man in a physical fight. Her excuse for aggravating the situation is that it is what any good mother will do and that also becomes the advise of the positive character. She knows she will outrage him blowing his week self esteem out of reality because now she takes him on as a man insulting him at the area he feels the weakest because he acts in the manner that he does because he knows from his weakness he can never manage to be a man in the company of men. The outrage she unleashes within him will satisfy the negative character. Reminding him of his coward ness and weakness will bring the monster in him to its full potential and that is exactly the plan. By her action he will be reminded of his weakness and that will be in response from advice given by the negative character as for the violence part and the shift in the blaming will be in responding to advice given by the neutral character. The child will now see how much she cares and that she does not hate but love the child to a point where she will sacrifice her life to protect the child. That will please the positive character. Then afterwards when the beating or molesting or whatever cruelty is completed she runs off with the male as the lioness did when the lion killed the cubs. The mere fact that she still remains with him runs parallel with the lioness accepting the animal behaviour because the human litter may not be dead but that is not her fault and in any event the fun can continue in full rage the next time around. Why call a halt to all the fun because everything will be alright after the husband pushes a few hundred dollars in the hand of the child or bay him a brand new whatever that will sooth all pain going around. He does not care for the chid as he can bay the child's silence. She does not care for the child because she allows the situation to continue regardless and the child does not care for the child because no one cares for the child in any case. After all the father did show remorse when he bought the child a new whatever. The theme is about money. If the father is caught he would most likely get a fine because the penal system does not encourage incarceration for such minor crime. He can bay his freedom even by penalty of payment and money wins the day.

In the event where the child stepped out of line and the father (or mother) comes down on the child as hard as he can, to shock the child with such force as to scare the child so much the child will fear any thought of repeating the incorrect dead ever again, the other parent holding the hatred will then come in and phone the police, contact the magistrate, get the executioner in preparedness and go on with all dignity going to madness. She will never allow such abuse. She will rather see him dead than punish her child. She will make such a fuss and such a scene that should anybody not get involved they will become assessors to crime and child brutality. This again is a charade where her characters enjoy every second of her instable behaviour urging her on to over react. Now the moment has arrived where she can show the child who is the devoted parent after all She can show the world just how much she cares for her poor little baby that only went about to destroy the child she was destroying in any event. She will never allow the destruction to stop because of an action a balanced parent sees fit. The law comes down so hard on this man as they wish to scare off any other parent that will ever try to correct the behaviour of a self destroying child. After all where will the next generation of criminals come from if there is a well balanced society and how many lawyers and judges may become jobless.

Every one in an influential position throws their weight behind the mother's instability destroying the caring parent for caring. The judge himself has three personalities to fight and no remorse but to uphold the law to the letter. The politician that helped creating the law realises there are many more unstable persons out there to vote than stable persons and with the majority being mad it is clear with whom he will side when writing the next law into the law books. And besides he also has this little fight going on inside himself. His negative character has an enormous advantage because to him was not only given one life to destroy but so many it can keep him busy as long as he can remain in office. His neutral side tells him to remain in office because that is such a lovely place to be and his positive character now with him in office can be the god he always new he was.

The madness runs deep as it runs wide leaving no pillar in any community in strength. The description I give may be mostly in exaggeration of the truth but the truth it is. Even if the slightest way of is applying or of finding such evidence that this is taking place it will show that with the least provocation the balances is shifting towards destruction and the way society is, is an indication pointing out more than just strongly that the human race is on the decline. The truth stands out as a sore thumb showing that there are massive problems waiting for man on his spiralling way down the devolution ladder.

This is part in all layers in society from the very rich to the very poor and every one puts the blame squarely on the others without accepting any blame. When this child becomes an adult the process not only continues but worsens. For the sake of argument let us make the child a grown up male. The child as an adult misses the mother that he had but also never had. From this he acquired the loss he felt with the loss he does not recognise. Now he is ready to find a mate. But being human we humans have the culture that mating is a life-time commitment.

The wife he is looking for must be someone like his mother but with a slight twitch. He hates his mother by now as much as she hated him all through his natural life. He confuses love and hate, compassion and punishment, caring and rejecting, in a way that allows him the freedom of becoming a most confused person. With all confusion running wild and seeing love and caring in the same light as not caring and running wild he goes on the prowl in search of a wife. He longs for stability, which he totally rejects. He wishes for companionship he finds to smother him. He hopes for security he does not care for. In everything there is a threat. What should bring devotion brings hatred because that is what he recognises but does not care for. Will this young man's three entities have a wonderful time. He came into the world unwanted because of free love, he was raised in anger because of a free and fair society, he was neglected because of democracy and now he wants to employ the culture that brought him destruction as a child. In loving his wife he hates his mother. In wishing for her companionship he fears for his life. In pleasing her he understands only rejection. He is even more bent on destruction than his parents were. Where there might have been someone that tried to show him rite from wrong he got that show of caring as massage that that is a person is out to get him. Any person alive including his wife that will make any effort to show him the wrongs in his ways he will reject to a point of committing violence. In that he will find the rejection he hates so much and he will kill to destroy that. He may or may not administrate violence when at home. That comes with the role of the dice.

The wife he loves he has to hate because to him that equates mother-love. To him loving someone is destroying that person. The one he cares for he has to destroy. That is the love he was taught to accept. This leads him on a journey of more self-destruction than any attempt his mother ever made. His yearning for the mother he never had pushes him on to every woman he can find. His wife being at home does not please him because she is not the one he was looking for. His morals are mingled like a mixture of concrete. The good and bad are so intertwined he cannot see light from darkness. He starts a life of adulterous affairs partly to destroy his wife whom he confuses with his mother and partly on the hunt for the mother he has a desperate need to find.

With devilish cunning this sets his three characters in motion where they can join forces and destroy at will. His positive character allows him to bestow the love he feels but cannot share onto the woman he is with just because he does not love her. His neutral character finds her acceptable because it will last but a night and the negative character helps with the charm because he will once more destroy himself and the woman at home, which as a matter off fact he truly and dearly loves. Unfortunately he does not recognise the love as love because he does not know the feeling of love. The love he recognises as love is the feeling he feels for the woman he is having the adulterous relation with because he hates her as she reminds him at that moment as the female figure representing his mother. If he really hated his wife he would be at home destroying her but as he loves her he does not whish to destroy her so he destroys her by not being at home. The positive character puts all his attention into charm that he throws on the female he is with. But because it is the positive character it also helps him realise that he can have fun in destroying the adulterate as the adulterate is the one destroying the one at home which he loves. The neutral character tells him to carry on because he has to punish the one at home for not being the mother he is searching for and the negative character is in heaven as every one around goes to hurt.

He believes that his wife at home truly loves him and for that he does not wish to hurt her but in the connotations he has about love he also knows that if she truly loves him she will try to destroy him. After all that is what love ones do when they show love. But because for the simple reason that she does not try to destroy him the neutral character holds that against her and make him believe that she is acting in such a way simply because she is not caring about him. Such behaviour stands totally in contrast to what he thinks love is, while his positive character keeps reassuring him of his wife's devotion therefore he should set his mind at rest as he will not lose her. The negative character finds this unacceptable and reminds him to do onto her before she can do onto him. The outcome is a vote of three to none in favour of the affair and another round of cheating starts for another night.

While the cheating takes place the positive character will remind him to love the one he is with as the one he loves is not there and the neutral character will tell him that as long as no one at home finds out then no one gets heart while the negative character tells him should his wife ever find out she is in any case getting what she deserves for not being the mother he wishes to destroy back. He will not enjoy her company and may not even enjoy her sex. He is in search of something and that something she will no be able to provide. During the relation he will get bored and then dismiss her as a dirty rag.

The worst that could happen to her is to let him find out that she has true feeling for him. That would place her in his power a place no girl will wish to find herself. The mother hatred will come to the forefront and he will start destroying her as he then can inflict all the injury he does not wish to inflict on his wife. By chastising her he will find some accomplishment and relieve and seeing her anguish will fill some of the need he finds in repaying his mother. But that will only bring some satisfaction and it will only last for short periods. But he will still yearn for his mother and in that there will still be a need to run more woman down. Because he does not have a clear image of what his mother was and what love is the characters can play mind games he will not understand. We all get some notions when one with a clear image of a mother and love between mother and child come across a female. We all have thoughts about what may be but we discipline the thoughts because we know we love the one at home and do not wish to sacrifice what we have for what may be. Hell, there is some woman I have met that is as attractive as any creature can wish to be but the very last thing I ever wish is to spend even one night with her. It is not because she is unattractive but to the contrary. She knows what she is and she knows how she excites men and she uses that charm to get men to dance around her with pleasing delight. The worst fate that can ever come to any man is to get involved with such a woman or even worse than death will be to marry her. She is the female of the male I described and she is bent on having men flirting with them and then just throws the verbal cold water on them to enjoy the reaction they get. In a marriage the first signs of trouble will drive her into the first bar where she will pick up the first victim and destroy her partner and the one she is with one blow just because she did not get her way in the argument she had with her husband. The poor slob that someday lands her as his wife will have enough information to write a book about hell.

With the conflict another situation with another disturbed child may provoke the complete opposing figure of which he is in search of.

In the next scenario of the Don Juan now in discussion holds the neutral character in place that will not have a clear picture of the woman of his dreams. With the image of his perfect woman being very vague any of the other two characters will come in and pour their versions of the perfect woman in his mind. The reference picture that he has about the woman he wants will be completely out of focus diluting his perception completely. He will want the woman every one desires. The disco queen or the brothel bitch or the bar tender out every night with another guy. He will wish to find the woman that treats him like dirt. Being treated this way will so kindly remind him of his mother and with his wish to please his mother as a child he will transfer that to his wife to be.

He will forever find some tart he wishes to please that has no wish to be pleased as she is in search of a man like her father. Her hopes are to find a man, and usually with success is one that beats the daylights out of her. Her first her second her third husbands will all have one thing in common. They will be woman beaters without exception. Or they will be drunkards, or womanisers, irresponsible persons but that is precisely what the woman has in mind although she will die before she admits it. If someone that truly loves her for what she can be to him turns up she will hate such a person because his devotion confirms her rejection. To her that man is the representation of her father and she in her twisted mind thinks her father was such a nice and devoted man because her characters will never allow her the opportunity to see her father for what he truly was. As her attraction and his attraction does not meet the requirement of their characters he will follow her like a lost puppy because her rejection of him is what tells him of her true love and devotion and that is precisely what she find so revolting about him. She wants a man that loathes her and here is a man that adores her. That throws her characters into disarray just as much as it throws his characters in disarray.

She hated her father as much as he loved his mother and because her father was the personification of brutality as was his mother they had to accept what they received as parents. But since neither had a real figure to relate to their characters turned the image they built around that parent around as to make them very acceptable. The characters they have will for the rest of their natural life bring to them the opposite of what they had as parent. That leads them to the very opposite of what they are in search of and what they find is what they wish for although it is exactly what they do not want. They both are condemned to one life of misery and if there ever is a hell, that place will be a merciful relief when they die compared to the life they have.

I do realise from the examples one must deduct that these cases are only the mental cases and anything more serious would find the person a patient in an asylum kept under lock and key by the President's special request ordered through the highest court in the land but it is not like that. The extreme I underlined because the extreme is the easiest to understand. But the destruction the parent has in mind for the child could be baying the child a very expensive pair of shoes only to let the child feel slightly important or giving money for a movie you would not wish your child to see but from the expectations your child has you cannot deny the child. It could be that you allow your child to go out with friends knowing the next day the child will write an important test. I do not wish to go into detail why that is part of the destruction or why it may be destroying the child because that is not the issue. The issue is the personalities lurking and being you. There is no slip of the tong with some wrong words slipping out. It is a deliberate intentional conveying of a massage to the other person about the true nature of your personal feeling and thoughts concerning the other person. What is the slip is your allowing one of the characters taking control for that split second while your guard was down. The turning of the cars steering wheel landing your mother-in-law in front of an oncoming vehicle while you were looking the other way and can swear under oath you never saw the oncoming car or had no inclination to go that way in any case. Why that happened is a total mystery because you will never do such a thing intentionally or other wise while the truth was that you were quite enjoying the nagging old witch's' company and her on going tormenting in her criticising you in the way you handling her precious daughter . That calling your employer by his first name when you were actually out to impress him. Calling the young girl in the office my lovely in the presence of the biggest gossiping bitch in town and realising this will lead to direct link involving a phone call by informing of your wife within seconds and knowing there will be hell to pay that evening while the truth is that you truly never even noticed how beautiful and smart and lovely and sexy and gorgeous this young girl was. Any thought of her appearance never crossed your mind for one tiny second. In another case your unintentional looking at a girls legs as she gets up and finds herself in a very embarrassing position for that split second while she is glaring at you for being such a dirty old man. Your looking down the blouse of a very breasted beautiful girl as she was bending over while your wife has caught you with your hand in the cookie jar. This is every day incidents with no intension of ill on your part but happened when you did not have full control of that situation for one instant. It all happened by accident but you believe me the deed was as deliberate and intentional as any of the cases I mentioned. It could even be as serious as your kicking a business competitor on the shin while you slipped and almost fell. It is the one character shouting your innocence as the other character is calling on your record always showing good manners and polite conduct while the third will never let an incident slip by.

This can and does even go as far as a nation. I am an Afrikaner Boer and being that I am not blind for my people's mistakes. When four Boere gets marooned on a deserted beach of a desolated island far from any other culture you can rest assured that within the hour of landing between the four there are, they would have started five different Christian denominations and six different political parties. It is a well established historical fact the during the Anglo Boer war at the battle of Ladysmith the Boere had twelve thousand generals with not one soldier amongst them. That is quite typical because we listen to God through His Word and no one ells. This last remark does not exclude our personal characters promoting twelve thousand times three different opinions.

....And then there are the Bible punchers...the ones that will convert you weather you need converting or not from ....well seeing that they never met you before it does not matter what faith you may hold, because only they can bring you absolution because they not only personify what ever they believe their god is but they see themselves as the direct extension of God, a finger or hand of God controlling life on earth.

They are hoity-toity, overbearing and haughtiness rolled in one god given container forming the BIG They. They can recite hundreds of Bible verses for minutes on end and that they believe is the key to their absolute presumptuous claim to God.

They walk with God and they talk with God and they discuss with God matters of mutual concern giving God advise where needed and as they see fit and where God is in their opinion straying from decisions taken at their previous meeting.

I am not referring to the normal God-fearing pious person that goes to Church and feels his thirst needs quenching on Sundays. I have no rite even to discuss any person's religious thoughts and belief. What I am referring to is religiosity to the extreme, a mental unstable drive of laying on the hands to heal, involve every one in religious debates every second of the day, starting to pray out loud as to draw attention of all persons around that should observe how their closeness to God has become as they are in constant prayer and will grant you some time between prayer because you should remember, they are keeping God on hold and the line is busy. Only they have the rite to prayer because after all when ever they get hold of you they wish to pray for you as if you have no connection to God, God has only given him a direct line and all others have to go via the switch board and wait their turn if they get a turn. Normally and in most cases, actually always I let them be but times arrive where they interfere so much you have to put some perspective in them. When I take them on issues their argument has the same logic as that of a pregnant pig and I have learned not to let them off easily. You press home the point and make them as big an idiot in front of as many as that wishes to listen, and destroy their mental thinking ability for the rest of the day. I have had situations where they tried to run away from me and I would run after them taking with me as big a crowd as I can possibly gather at that moment and destroy their image to shit. After such a session they are normally so annoyed with me they ignore me flat where I then leave them alone. I would never do that to other people for no reason can be important enough to humiliate your fellow man. But with them they leave you no other choice.

Not once and I repeat not once in my life did one person ever come to me with the introduction of: I come to you in the Name Of the Lord, and that bastard did not cheat, swindle, rob me, or steal from me. I have reached a point where I decided that should any one in the future introduce himself and refer to "I come to you in the name of the Lord" I'll chase him from my property like a bad dog. They are the biggest crooks and con persons walking on earth.

They too have a massive trio rage where their characters use religion to go out of control knowing very well all people will respect God and their referring to God always bring the other person in obedience. That is when they hit home with the most devious cheating and swindling you can imagine. If someone of their likeliness offers to pray for you don't close your eyes, grab your purse!

They're knowledge of the Bible is astounding but they know nothing about the Bible. They learnt a thousand or two thousand texts and recite them with speed, throwing the one after the other without making sense or allow the meaning having any connection. That is only an eye blinder, a way to astonish you so that you will lower your guard and that is when they hit home. There is no such a thing as a free ride and they always want from you ten times more than what you are prepared to give and when finished with you in the very last paragraph of the small print area it is only all about money but of course "in the Name of the Lord".

You wish to spread a gossip or any untruth, well be sure to use their channel. Normally those services they provide for free but then it must be juicy, unfounded and completely void from truth. Their trio works on the basis that the positive character takes them to personify God, therefore they can do no wrong in the eyes of the Lord. The neutral character advances the notion that they may convert you and that may be useful in some future schemes where you then can fit into more devious plans while the negative character is of the opinion that by your not believing the way they do, you are doomed in any way so robbing the convicted bears no shame. God put soles like you on earth to be useful to their likes where after you will go to hell anyway so what the hell, they might turn some profit before you meet your final demise. After all if you cannot see the light they give you, you may as well be blind and being blind you don't need more than what you can see. By them taking from you and giving onto them they are receiving with a self help scheme what the hand of God on earth deserves and where you are going to lose everything by your departing to hell you might as well start at a point where you are still useful to their blood sucking. Should any reader not believe me take some time and start discussing non religious issues with intent on your part to learn some angles their characters maintain as informed opinions. They are not hard to find. As with all criminal hoodlums hanging out at places of criminal conspiracy they normally hang around at the tents of the evangelistic preachers commonly referred to as the "Happy Clappies".

When I shared time in clinics with persons having some psychological problems this was the beginning of my theorising in this direction. I wish to state once more it is merely an observation of a layman fighting to analyse his own condition and took time to see where similarities were between different humans in the same boat that was sharing a mutual difficulty. Some were alcoholics where I am a teetotaller but still behind the condition I came upon similar causes giving one person one crutch and another person another crutch but it is the crutch one has to loose and behind the crutch you have to find the pain causing the person to grab for the crutch. What ever I share must be taken as not even an informed opinion but as merely another opinion where we all have opinions and it is worth the while to share opinions of an assortment and a variety.

In the last part I indicated persons holding the righteous views about their religion to advance and use as an excuse for their almost and sometimes definite criminal behaviour and malice intent. On the other hand I have seen suffering where these characters dish out what no one can bear. The agony and torment some people go through is of a much higher pain than I ever suffered when I came off my bike. The pain is more real than physical, the fear is stronger than death, and the confusion is louder than not understanding. It is horrible because some of them feel the anguish moreover than they would if a genuine murderer was chasing them. With a genuine murderer you can try and escape or hide but in their suffering there are no such luxuries.

The way they suffered and reality of their hallucinations had put the fear of God into me and made me more than willing to get over whatever small difficulty I had because my luck is worse than the Irish. (To my mind no one can have worse luck than the Irish because they got themselves in a spot on earth from all the places they could they chose their spot next to Brits, where the forever meddling and interfering bossy Brits is occupying the very next island) With such bad luck going my way I might just find my problems increasing. I had electro convulsion treatment on several occasions and to my opinion the treatment has a healing affect as the neutral character loses some dominance with the loss of memory through the electric flow. The patient then loses confusion by gaining perspective where the dominance of the entities reduces. As the generating of electric flow brought about by the life factor decreases electric tension in the brain the location of the entities become affected as well

Gravity, electricity, time is all the same thing and in a more or lesser manner influence life and moreover life in the brain. By reducing the electric tension the mind stimulates and as the convulsions allow electricity to escape from the sells stabilising the brain activity and helps sorting the influence the characters have on the person. I must admit that directly after such treatment I don't feel such an excessive urge to speed. Fortunately the condition normalises quickly and I can get to enjoy the exhilaration of my crutch once more. That proved that influences of such characters do vary and can be in dimensions of interpreting and it seems that influences from outside sources can be a major consideration

You may believe it or not considering my poor academic background but I have an inquisitive nature and a need for mental stimulation by acquiring facts and information and that was my academic downfall. My positive character always urges me to test another person's knowledge base and interpretations of facts. This was present even as a scholar as I did forever test my teachers. My neutral character will then classify him in filing order typifying and classifying ranging from brilliant to shit where my negative character will fore ever test the teacher in relation to my personal abilities and from that stance supply the necessary admiration or animosity. I never allowed my teachers at school to escape and when I became suspicious of their depth of knowledge hell would be upon us, moreover on me because they had the cane and always knew how to use the thing. However in cases where the teacher became a source to quench my thirst for knowledge I would eat from his hand. My positive character would take charge and push the others to a silence where no one knew they existed but in the other events of me growing suspicious about the teachers abilities the negative character would destroy any form of harmony that may develop between us and that was the normal in all but a few cases.

In primary school the teachers thought my behaviour was cute but in high school it became intolerable for all parties concerned. I make this remark to indicate that outside influences does play a part in younger minds and through positive stimulation the influences on the characters can be directed to a positive outcome for the child. No one lives in a tight cased cement container never to have an ability for change. It is the duty of the teacher to recognise and direct the children's interests to the benefit of the child and that could lead to the benefit of the class.

I established my theory with all relevant information based on my personal case. While this was going on I also realise there are nothing about me being exceptional or unique and if this applied to me so would it then apply in other cases. With a clear objective I started discussing other people's situation with them to draw similarities that would match my case. It was similarities I was after and not parallels so every time I got behind the whole issue by befriending the person and in that way I could establish a confidence that no professional could. As there was no malice intended on my part and I did not brief the person on my theory or tried to offer remunerable advice I could see no harm coming to anybody. The information was never brought to paper establishing personal files and since gossiping is not one of crutches no information of any private nature slipped past me. If ever any advice came from me it was certainly not on the grounds of my theory so I could not harm any person in any way.

But the more I came involved with other persons the lesser the importance of my personal issue became and the more I detected some golden thread running along lines undetected. In some people I could even detect which character reigned supreme that day by remarks they made or the moods they had and in limited cases there was differences in facial muscles as the mood swings occurred. But in the Afrikaans book which up to now a very limited number of readers had access to, this is the first time I went as far as mentioning my observations to any person. I can even remember the precise moment the light of understanding went up when a psychiatrist Dr. Steenkamp, which is still treating me explained about the mind and the free will of persons' personal thought, the way a person react on they're dissensions and the total absence there are of demons or other spirits that may influence the mind. I state this categorically I have never commented about this theory I have and least of all to dr. Steenkamp so I do not wish for any person to conclude he had any personal opinion about my conclusions. I merely said this because only a few incidents stand out in my life as very memorable and this was one such moment.

What I found was that it was as good as a human trademark apparent in every one, slightly more apparent in some than in others, they are in every one all the same. The entities are mostly absent but come in when a person has his or her guard down or when a person has emotions with an influence stronger than the person's ability to control. All actions man make is with intent. Some might be under the guidance of one or more of the characters but every one has full control over all their deeds, without having an excuse for conduct. They rule your life as much as they are you and will promote your true intensions when ever you do not wish to. Fighting the characters is fighting yourself but you can and you must find a way to recognise them because their intentions are to harm and never to uphold. The characters come with certain emotions bringing along certain feelings. The feeling may be an excitement that does not match the situation or an anger that does not fit the occasion but if one is vigilant you may catch the feeling before the feeling catches you. It may be very slight in irritating handling like for ever pressing the wrong "t" being the "y" on the keyboard or turning a cup of tee over on some important guest or having a dislike in someone you never met before and should not have any special opinion about.

One incident to try and prove my point I wish to raise is from my personal recollection and I wish to share it in order to avoid other peoples' affairs. I suffer severely from acrophobia. Putting me on a double-decker bus is about as high as I can go. In all sanity taken seriously one cannot have acrophobia. One cannot be afraid of heights when steel bars inches thick will prevent your falling. One cannot be scared to look down a glass window when it is closed and you cannot fall through. It does not make sense and yet I can assure you it is a fear greater than the mind itself.

The fear is irrational but should any one try to loosen my grip once I grab onto something he is not only endangering my life but he is seriously messing with his own life. I am aware of the problem and believe I do have a rational mind until it comes to heights. There is no thought, there is no reason and there is no arguing about matters. It is instinctive irrational animal-like behaviour where I go into a survival mode. I cannot fly and yet I know flying is the safest form of transport so much in fact it may be a thousand time safer than my cars or bike, but that is the rational and it disappear when I look down and see something small down there realising it should be big. Even just the realising that I am about to leave the earth is more than I can control.
The fear is almost if not an obsession and becomes uncontrollable. Then one day i stood on an exceptional tall building (well exceptional tall for me a person coming from Ellisras the true one horse town in the middle of a semi desert) of about eight to ten storeys high. I did it purposely to see what the emotions was accompanying the acrophobia because the attack must have some prelude. It can't just hit you like a brick that is nonsense. Something has to form a fore play, a sign of what is coming. Even if it takes one second it still is there. Nothing can just overpower the mind instantaneously but every thing must be about a collective of factors and facts coming together.

Coming out of the lift as I was walking towards the corridor where after entering it I could look down for the first time I intentionally was waiting for what ever to come first and announce the shock. Then as I came to the open I felt the feeling of fear but it was first another fear, one I was use to and knew. It was one of the characters coming to the foreground as if called. That made me realise that it was not the heights I feared but the negative character. I feared the character might take control and make me do what I did not wish to do. The fear of the heights is there, and that is no maybe but that is an extension of the problem and not the problem as such I may not fear the character and I may feel uneasy about the heights and it could even be that I become insecure but the problem was outright the negative character coming to the front.

As my dominant personality losses confidence the negative character comes in. It is not a case of him pushing me or my jumping but it is something going on in the realms of my mind where I do not understand all things all the time. It was a fear of what I may do to myself and not of the heights. Under all of this was the presence of the negative character lingering almost like a shadow feeling not present and not absent but just there. Then came the shock of the actual height and all logic flew away like a little bird. I was clinging and grasping for dear life.

If I can take my mind back as far as I can go back my very first recollection I can recall is a scene where I was on this huge tractor and it was far down. I was definitely under two years of age because I know which farm it was and my Grand Father sold the farm in the Free State that had the tractors before my second birthday. When we moved to Tzaneen he farmed without tractors so it was definitely before my second birthday. I was on this enormous tractor (enormous because of my youth) shouting desperately as I was crying hysterically for help because I remember the thought that I had no chance of getting off that tractor all by myself. I do not know how I got off or who helped me off and being where I am now I must have gotten of because I am not on the tractor any more but that day my negative character and my acrophobia met and got mates. Of that I am sure and if I am correct, then parents should take care not to scare their children in an innocent prank of unintentional fun with their child's fear. It could have lasting consequences to the child. It is not the heights I fear but it is the character taking control even if I know there is no chance of that happening still there is no logic as far as the phobia holds ground. If that is the case with me it should be the norm. People are not scared of objects because objects hold no threat and every one knows that.

My humble opinion is that one of the characters is dominant and the dominance is so much that the person feels threatened by that character. The character is in control so often through a depression or an anxiety or a mania of sorts that when situations arrive where the scene should be normal fear becomes the norm.

The negative character brings the threat the neutral character bring the warning about dangers and the positive character joins in by bringing the fear. The positive character brings the fear in to dislodge any attach the negative character may launch. The positive character on the advise of the neutral character disables the body and disables the ferocity and fierceness of the negative's dominance. By pumping adrenalin the body goes numb, the legs go weak, the arms shiver and the body has no strength to function while the person who is in the middle of the junior civil war see object as the reason for the attach but the object is only the trigger and not the show.

I too, am of the opinion that these characters and the way they perform their balance on the day and in the situation makes the hero or makes the coward, depending on the balance at that precise moment and occasion. Phobias connect to the negative character and mania to the positive character and by allowing un- protection through some situation triggering the mini civil war inside the mind; the person becomes a bystander where the person should be the controller.

A kleptomaniac may put something small in a bag and swear by the fact the kleptomaniac did not know about the action. That may be as much the truth as it is a lying because the actions was not deliberate, but the actions were intentional. It is easy to ignore the compulsive behaviour and claim non-participation when participation may have been semi unintentional but still enjoyable. The one character may distract the attention of the person but it is done with full participation because in the end responsibility is with the person. One would often hear the remark: "I knew it I knew it was going to happen" when something was going wrong. All humans can read situations and your mind told you something in the situation were desperately wrong. While the positive character does warn you of events coming it is a deliberate action to allow the neutral character to distract you while the negative character can play for time for whatever occurrence to take place. The whole scene was a deliberate action by the person to gain a negative outcome to produce some suffering or hard ship to some degree because that is why we are on earth. But I shall get to this last remark later on.

This takes us back to the bleeding heart baying off guilt by paying the philanthropist to collect on behalf of the Hoggenheimers dishing out to the Mammonites paying the Mammonists for some slave driving. The actions are deliberate but the true intentions are deliberately unintentional. We are bullshitting our conscience for gaining our mistrust. It is the lye of culture and all participate but some participate to a degree that does not please others. The degree might be to some extend not serious enough to bring commitment and the persons would stand on the side line and criticize without direct involvement because of fear of own guilt uncovering or even of a want to participate while others would come in and rescue but not to save but out of spite because of personal yearning for participation that the person knows would not be permissible.

Another scenario is where a person is drowning. The rescuer comes to save the drowning victim. The victim is exhausted beyond normal mind control. The lifeguard reaches the victim whereupon the victim tries to drown the lifeguard. The victim has lost rational thinking and is then in a mode of action versus reaction. The positive character grabs and clutches at the lifeguard in anticipation while the neutral character tries to survive the conscious and the negative character wants to save the situation by taking the lifeguard with the drowning effort because after all it is the responsibility of the negative character to destroy and destruct as much as possible.

Some gave these characters names. The negative entity goes by the name of a death wish. The neutral character goes by the name of don't care. The positive character goes by the name of optimism. By naming them we found once more a way of avoiding our duty to recognise. It is much easier to dismiss than to admit because admitting has to lead to prevention and prevention is no favourable option.

We say the drowning person got panicky but that is another word we use to escape from reality as much as an excuse in avoiding responsibility. By being panicky we deliberately excuse behaviour and responsibility about actions we may commit to explain irrational behaviour. The negative character wants to punish the lifeguard and even make him pay for his life in interfering with a situation the negative character is enjoying thorough rely while we others only recognise the efforts of the positive character because that will be the nice thing to do. The next time we behave in the manner as to kill our saviour we too can be acquitted on grounds of incompetence. The action of trying to kill the lifeguard is as intentional as the trying to grab onto him to be saved and as intentional as loosing control through the neutral character.

When I first read about the Lovejoy syndrome it brought to my attention that not all mortised control of the body is in the domain of the person all the time and sometimes there are some part of your life that can take control of your actions when you are not in absolute control. It is all a mind game you play with yourself in diverting responsibility with the compensation of enjoyment but the avoiding of dismay about yourself. There are no excuses because the final responsibility is in your power. It is in your power and much more even it is your birth duty to fight the characters but the moments you lose the fight you take the responsibility for the actions because you momentarily lost the fight. It is a win lose situation where only you walk away with the prize as much as the punishment. The muscles are under your control and you are in charge but the entities are little pests being you and are thorough testing you by grabbing control whenever they can. The entities are not only and absolutely negative but are positive as well. We have all been through situations where we admit afterwards we came through by the grace of God. We always hear some one remark that he or she does not know how they did "it" but "it" came through far better than "it" should under "normal" circumstances.

All our phobia, all our desires and all our hopes in achievements we pin on luck or the role of the dice but luck and the role of the dice has nothing to do with it because or achievement good or bad as our accomplishments wrong or correct, and our thoughts being acceptable or not is within us, in our control as much as it is us.

A paedophile should be hanged from the nearest tree because he gave in to the want of the characters and not because "he is not in control of his actions". He wishes for the characters to take charge and even deliberately set up situations where they may take charge, because he enjoys the dead as much as they do and more because they are he. They are in all of us but for some certain behaviour are unacceptable and for others it is not. The judge bringing judgement knows in his grain that the molester will commit again yet he sentences the criminal to a few years of incarceration and is fully aware that their is no chance of rehabilitation because the culprit does not wish to be rehabilitated. He will find his rehabilitation as a death sentence because molesting children is keeping him alive. When the judge do not bring the death penalty he, as much as the culprit participates in the next cycle and therefore must take responsibility and participation in the next round of child abuse. But the judge sits there with the idea " there am I but for the grace of God", which is true in a way but also is not true in a far bigger way. With my fast driving I do not wish for a cure, but I know something is going to go wrong somewhere some day. That is my chance I have to take. He upholds the same argument and goes to the molesting because the sentence is the chance he has to take, and is the chance he takes. You can bet your bottom dollar that should I know before hand the next road race would kill me I would not participate, and the same goes for the molester. With all certainty that the hangman's loose is waiting he will have second thoughts about his next molesting session. Murders always fight for their life by fighting the death penalty. The underlining is that there is no black white and grey. There are no clear-cut defining borders and sides. The neutral character can be as destructive as the negative character can bring a positive out come. When the mob comes out to lynch the actions of the mob are negative in lynching but as they do not wish the continuing of the criminal's behaviour it becomes positive when doing the demonstration. That it will bring conflict to the child's guilty feeling and in that sense their actions are neutral to the child, the participation is positive in preventing the repeat and there by positive in the negativity of lynching where the police prevention is negative by protecting the paedophile and that is positive by upholding justice as it is neutral by delaying another criminal's relapse in crime because relapse he will.

The atheist does not go on a disgraceful child molesting campaign because he thinks there is no God and if he does not get caught there is no punishment. He avoids indecency because he is human. The paedophile may be the biggest Christian around but argues that since he is only human and humans are sinners and sins are alike he can maintain his behaviour until judgement day. Such a line of argument is very typical of the trio being in charge. The responsibility is always with someone or something else and the person never pin it to specifics but shifts the blame to wherever is convenient. And so does the Judge! Hang the bastard and judgement day to him the sinner will come a little sooner. If he does not wish to control his characters help him by eliminating him with his characters. By molesting the child he starts another cycle producing one more child having little control over the characters the child will fight when he is an adult. Stop the violence by stopping the cycle by stopping the one not controlling his characters. If death waits as a surety he will mend his ways or seek help to accomplish change. It is easy not to change and difficult to change but we all can change.

One may have be opinion what I refer too is about the good and the bad, about the saint on the one shoulder and the devil on the other and think well…yea I've heard that before… But it is much more than that. People stop at serious motor crash sites not with the intension to help. If any one admits to that that person is untruthful. They stop to feed the urge. People watch blood sport, not for entertainment but to feed urges hidden deep within the mind. When an armed robbery takes place with possible killing spectators run to the scene. When there is a fight on the schoolyard the word spreads like fire and little else can generate more enthusiasm. The most brutal serial killer always receives the most male. Woman would throw themselves at the criminal misfits with marriage proposals; coming up with the excuse they have enough love to concur the beast's evils, but that is a hideous lie and they more than any one else know that.

They wish to share in the darkness of evil, find someone that could lift the veil covering the beast within. Stopping at the scene of a bloody accident evokes prime animal senses covered by social upgrading but is still very much lingering within. It serves as a reminder about the days when a feast came from such human blood. At some time all of man was a cannibal, man-eating monsters that feasted on the flesh of the enemy after a conquering battle brought victory to some and victimisation to the others. When faced with starvation it is a normal sense that kicks in where people will start eating human flesh. Shocking, as it is… the shock is about realising the urges more than revolting.

We all have the darkest desires of committing unspeakable atrocities and most barbaric acts that will shock the normal mind into panic and frenzy. Some people dislike blood sport, not for what they're purist of heart tells them to reject but the rejection comes from the craving they fear. They fear the need for such beastly acts may linger and run out of control. A story about a mass murderer is best- seller weather it is fiction or truth it is popular. It sells and there is a valid reason for that although our minds reject the reason.

When thinking about these crimes we put it in the basket of the negative because it is where it belongs. No one can ever be positive about such behaviour and with that we create fabrication of truths we wish. It is as negative as it is positive...it is neutral. The good / bad character is not the danger; it is the neutral that is dangerous. It is the neutral taking our minds on fantasy journeys. It is the neutral character we most easily identify with and mingle with. It is the neutral character setting our morals. When Ted Bundy killed the many woman that he did with the brutality that he applied he was not negative, he was super positive. To the woman "forgetting " to lock her door, walk alone down the dark ally at night, stopping to give a strange man a lift or hitching a ride the deed is not negative. They are super positive in their expectations of what may come to them. Sure, they do not ask to be raped, or beaten, they do not beg to fall victim to crime, but lower down in their minds they know of such a possibility and very, very deep, deep down they stand neutral to such an event because they do not mind the excitement or the sympathy afterwards. The bank robber, the pickpocket, the shoplifter, they are not negative they are super positive. It is being neutral that is the negativity.

The animal cannot see the universe from any other position than the one the animal holds. Singularity placed divinity in the centre of his universe and the animal has no means or brainpower to translate his mind to another position other than his centre where his needs and desires are. From there comes the neutral holding positive or negative as well as the he. Negative means he should fear and positive means the other should fear. In this religion can be one form of atheism and atheism another form of religion. Atheism is the inability to see anything except from the centre of that person's universe and the inability to transform to another position seeing it from another viewpoint other than that persons universe centre. When Fish sent the parents of the child he cannibalised a letter informing them how he devoured their child he was positive and the deliberate pain he caused was positive. Peter Kurten the vampire of Dusseldorf only became neutral with his last victim, but he always remained positive to his wife so much that in the end he forced her to betray him so that she could claim the reward. By her collecting the bounty he realised his final act of being positive as he always felt about her...but only to her and so his final offer he gave her was that the bounty that went her way. To him that deed was as positive as his killing and eating of three children in one night was positive because he was unable to translate his centre point to their position. That is being animal in every sense and animal on two legs. The Boston strangler was always neutral working his way to becoming positive for days although that meant the utmost negative to his victims. Being human and religious is the understanding of other concepts forming a centre way beyond your centre of the universe and moving away from the animal neutral to a human stance Not eating the flesh of the sheep does not in any way make you more positive but it makes you too stupid to see the universal picture of the totality of Creation and its wider meaning. That is trying to prove you are not the animal you know you are but cannot seem to separate from. In the neutral character is the one that no one propagates because that character is much to close to us to be comfortable with, that is the animal, the beast or civil the morally accepted or rejected but still nursed by all because of pre man mentality. From that the other characters stands positive or negative but the neutral character holds the centre and the key to light and darkness in the mind of man becoming man. Claiming the position all other characters including the self occupies a part of the mind. The neutral character sets the tone and the rest will follow. In the neutral character lurks the animal and depending on the person, the darkness of the animal comes to the foreground or stays in the back ground but is forever present in the mind. Identifying with or identifying the neutral character places us in the realms of man or animal. Setting us apart from the neutral character is what produces man and not animal. In the same manner as not eating sheep is the saving of one criminal life where every one involved knows that criminal is destroying dozens of future lives and expanding the problem by creating many future criminals where they then would create some more criminals at a ratio of a dozen to one. In three generations the one criminal is then the cause of hundreds paedophiles walking the earth. That is degenerating civilisation and it is all because the law enforcement from politicians and judges through the lawyers and civil servants down to the bleeding heart and the cop on the beat that wish to prove them not being the animal that the criminal is. By creating the environment and breeding ground for the animal and forcing their positive ness in neutrality they destroy the future of the following generations of man. Law not taking blame for their actions in the neutral stance is as much being a criminal as the criminal's neutrality creating his positive to negative relation. The law enforcement' officials excuse for not wanting to be as bad as the criminal is more destructive and a much bigger misdeed to society than what they do to the criminal because after all it was the criminals free choice to commit or not commit and as much criminal not to wish to pay for his deeds. Either we teach the criminal there are billions of centres to the universe or remove him from our centre but allowing him to become forcefully our centre proves as little as it is destructive to every body. But by throwing him in prison where he shares time with others the same as he accomplishes only that law enforcement may promote crime to establish more cycles for their and all the other criminal's benefit but to the disadvantage of man in general.

The paedophile or any other criminal or anti social behaviour is quite rectifiable where psychologists must teach the criminal in recognising the incorrectness of behaviour, the recognising of the characters and the control of the characters by thought control. Criminality starts with a thought and that is the point I started in my discovery of the characters. Being suicidal as I am and that being my problem starts with a thought starting with a feeling leading to a depression. I am never suicidal on my bike or my car doing high speeds. Then there is only the positive character with the neutral character keeping guard for traffic control. Never do I at any stage exceed my personal limits or endanger lives through recklessness by outsmarting my personal ability where my mind works at the speed matching my vehicle and that is the secret to success. It starts with a thought. It starts with a feeling. That is the gate to progress the characters follow. The remedy is striking a link where one will recognise the thought sparking the feeling and recognising the felling sparking the thought. There is countering the feeling by thought as much as suppressing the thought by feeling but the BIG issue is that you HAVE to know yourself. The alcoholic feels the urge for booze as much as the thought for liquor but when recognising it he counter acts and that is where alcoholic anonymous has such a great success. If he did not wish to recognise the urge as much as he did not wish to suppress the thought alcoholic anonymous can do little. It is the difference between cure and tramping. But it is fighting day and night and fighting yourself knowing you are in the fight for your life for the rest of your life. It is fight you can never win and must never lose. It is continuous round after round with no victory and no defeat, no prizes and no glory. Only shame to follow defeat and the winning part has no recognition for effort or accomplishment but to you. That alcoholic fighting himself and finding his determination is the human victor. To me staying neutral is staying alive. That is what prison should teach the criminal in recognising thought and control thereof. The big courtroom confessions these murderers make are about showing the world how big their universe are, its about how they are the centre of other's universe and the control they had in destroying the others with their universe but it is never about apologizing to show they can understand what others have in the universe they destroyed or that in fact are others with an own universe to have. Should he think he can outsmart be unable to rehabilitate he is animal and should become destroyed like all other raging animals being out of control. He should therefore fear death or find death. In the modern penal system rehabilitation is a word used in the courtroom with no other place to have. Life should be about improving and not sustaining.

As long as doctors treat suicide as a state of being morbid or depressed they will not get anywhere and it will take them a long time never to come to any cure fort his state of mind.

Alcoholics and all drug addicts lose the neutral character when intoxicated. The person becomes super positive loving the world for the world loves him right back or becomes super negative by getting aggressive or drunk with remorse hating his parents for what they did or never did. In this the addiction of the parents always plays a part when the child of the alcoholic also becomes an alcoholic. When the addicted becomes sober the neutral character takes control with vengeance, as the addict needs the next round of substance. To the neutral character surviving means finding the next round of intoxication. There is never a balance and when the positive becomes positive, drunk or sober, it shares a spot with the negative and of course the other way around also applies. The drunkard is prone to mood swings that is apparently out of his control, but that is a fairy tale. When sober the same applies, as his moodiness seems to follow his every move. The drunkard uses the substance to escape from himself and his own adequacies he feel about him in his self-portrait and therefore allows the characters a free hand when intoxicated. This we in South Africa call Dutch courage and will have different names in different regions but all names apply to the same behaviour.

Things are normally not as serious as in the case of the child molester or the mass murderer turned cannibal. On average it does not have such deep and intense underlying emotions and fights to the bitter end. It could be a case of the young man is meeting his in laws to be for the first time. The dinner is in great preparation as it is in great anticipation for all involved. Every one but most of all the lover-boy and soon to be in-law is under pressure to impress. He is the new face in the family having all eyes on him because every one knows all the other faces and he only knows his face, which he cannot see in any case. He is out to impress and is tuned for this all out effort of do or die. In this effort he relies without relying on some help from the characters because he can use all the help he finds. He lowers his guard completely to allow the positive character unhindered passage in the situation arriving. In the background lurks another character that is far less dominant in the whole affair.

It will take but a flash in a moment for the negative character to prove a point but as he opened all the doors so wide for the positive character he now is even positive in being neutral. With all the tail wagging he has to go through his negative character takes little appreciation in any possible discomfort on his part. Lover-boy is yearning for acceptance to such a degree this making of his all out effort is rather new and strange for the characters as they show an all out help line helping by the full range of their individual abilities and to use the chance in such an unhindered open channel participation. But one is waiting his chance in quiet anticipation. At the moment of climax where father in law to be wishes to make a toast will be the moment all three characters join the fun by creating the incident all three have been waiting for all night. It will be to the determent of boy-impressing of course. The negative character is very aware that the neutral character may prevent his planned action there fore the strike is lightning quickly. The neutral character sits lapping up all the attention as the positive character is all out helping with the impressing of all around. With every one well occupied the chance comes for the negative character to make his point for the night since everybody was having fun and he had to sit idle and unwanted.

As the tray carrying the wine glasses, which are filled to the top, passes by lover boy, the negative character motorizes the arm closest to the tray and hits it with force. The action holds the speed of a boxers punch and with that lightning speed lover-boy never thought he had such quick muscle movement. The action reaction reflex action reflex reactions is beyond the abilities our boy to be married ever dreamed he had. He is totally surprised at his quick ability in movement and that to his knowledge is miles ahead of his ability to move. In all his life his arm never moved that fast as he turns the tray over on the lily-white table linen.

With him being totally out of place and out of sorts he cannot recognise his ability and that is what his positive character ensures him. His neutral character will be very embarrassed and that embarrassment he places in lover-boy's private embarrassment about what happened. Shouting and proclaiming accident by him as well as the other two characters will bring all members of the in-laws-to-be under the impression that this was indeed an unfortunate incident completely convinced by his very genuine embarrassment. The embarrassment he proclaims are partly his but mostly the embarrassment comes from the other characters proclaiming innocence to him about their involvement and misuse of trust. This he uses to further his embarrassment to exclude all blame of deliberate action on his part in order to impress his in-laws-to-be.

Every one present will feel deeply sorry for him except his negative character that made the point that all the licking may be for tonight but he is still his own man and will have is own way in the future to come. The question is was his innocence really that innocent and was his embarrassment truly that hearty? On the first count no, he was just as deliberate in the overturning of the tray as all his other actions was though out the night and innocent is the last thing he can be guilty of. On the second count, well yes, in a way but not only for all the reasons he proclaims. No one can ever proclaim innocence and non-participation in deeds the person commits. All your actions are all your responsibility. No excuses can be maid without lying through your teeth. Being born means the fight is on and the fight will continue till your last breath is wind. Having trinity around is your birthright and fighting is the option you made before birth and not after birth.

Life of man is about fighting yourself with all the vigilance you can ever muster. That is why you are here having the time of your life for all your natural life. It was your inheritance the day you were born a human. It is not all about negativity but it is all about achievement. It could be quit within your self and it could be in the presence of thousand of spectators. Every sportsman has "on" and "off" days and there is such strong emphasis on the physical that by training the physical no one notices it is the spiritual in training. The spiritual always is the physiological but all preparations are psychological. Practising is about training life to manipulate space-time to the best affect. In all it is only about the physiological. Training is about telling the muscles to obey command and telling command not to obey the muscles. When the muscles shout in agony to stop, command must turn a deaf ear and when command shout to the muscles to go on the muscles must be like a dog and react without questioning or arguing. To describe this we use the name fitness. It is the conditioning of the flesh but it is much more about the mind having control over the body and all fitness is about life commanding the structure in occupation to do what life wishes to be done. Fitness is moreover about controlling the characters than the body. You have to control the negative character's destructiveness to be about the opponent and not about you. You have to control the neutral character to dismiss outside interfering with your efforts. You have to control the positive character to bring subduing confidence. You have to use the fear to your advantage in believing you are fighting for your life and not merely a trophy.

Every sportsman has a story of "absolute brilliance" and always the remark is about the sportsman not truly believing he had the ability in accomplishing what he did. When saying that the sportsman only considers the physical aspect and never the mental drive that pushed him beyond the limits he accepted. On another occasions it is the very opposite when the sportsman declares the day as a disaster because notwithstanding an all-out physical effort nothing went according to plan. His muscles did not respond, his legs were stiff; his arms were not in synchronisation or what ever the excuse for the disaster is. When shove comes to push it is the three characters we find again behind the success or disasters.

The apparentness comes through by him not admitting his efforts links directly to his state of mind and the blame goes to external factors. Everything went just rite or just wrong. It is everything that holds the responsibility for his success or his failure. It is external forces at work and in charge of his luck or bad luck. With that remark he indicates his absence in his actions. His lack of admitting participation and his unwillingness to claim success or admit failure proves that he is relying on something he feels he has little control over.

We all admit that when we are tired we make mistakes. When we lose concentration things go wrong. When being absent-minded we make accidents. That is admitting to the role the characters play. Being tired means relinquishing control, letting go and then the character take control. But also when in fear of one's life you find yourself in super control where you are miles better than ever. That too is the characters at work. It depends on which character takes charge and to what degree does the character take charge. I always teach my sons never to fight a man in front of his girl because you face a man much better than normal. If you are winning a fight leave the other person a way out to escape and never allow the impression the person has no way out. As soon as the opponent gets the idea he is fighting for survival, or he has to fight for position in the tribe in example for the favour of a female you have a monster on your hands and in that case be prepared to fight for your own life. Never allow the idea to enter the opponent's mind he has to win or ells... that will be to your determent. In the instance where you do not allow an escape route or a man thinks he is fighting for a female you have a raging bull and three characters to fight and you do not wish a fight him with his characters on his side fighting against you. That will be your death you wish upon yourself.

It is not schizophrenia I am referring too. I am no psychiatrist but as a complete novice I do not believe in schizophrenia. I do not believe the person is hearing voices from beyond because there can be no voice of beyond. It is his imagination and his characters playing mind games and if a psychiatrist or a psychologist come in and join the fun admitting to such a scenario. With such outside help and sympathy to help the dreaming along the characters then can and will come out to play. The "split personality" in the fight is within every body and can go into rage whenever allowed to do so by deliberate actions, provocation by other, tiredness, fear but also delusions, that is true.

From the article so far one may tend to get the impression it is about big issues like meeting you're in laws the first time but it is not. It is every day all the time things. Sitting in traffic waiting for the light to turn. The positive character slowly shifts to neutral and the neutral slowly joins force with the negative that was in the background all day. Without the person noticing the whole situation within him shifted in frame but he is unaware of it. Suddenly an explosion fitting a war burst to life. Another motorist sits daydreaming with his neutral taking him on long trips because his positive has nothing to do and his negative is in the background. The light turns green and the daydreamer is just that tad slow in responding because he was in thought. He was in thought as much as he was on Jupiter. His negative character helped the neutral create a situation the negative character can participate and not be bored. So his slow acting is as deliberate as his breathing is and the other person with the shifting emotion sees this, recognises the stunt that the daydreaming motorist is pulling and with the positive character standing on neutral ground and neutral fully in the negative territory, the negative character comes in with a punch like none. Suddenly two very timed persons go into a rage because of the smallest incident. They do not recognise their own behaving and if they do not get their characters on a leach quickly the characters of both men will take charge and blood might flow. This we call traffic rage. It is a very convenient name.

The man sees some one he may regard as rather attractive and smiles at her. The positive character becomes embroiled in the proceeding while the neutral character joins by taking control and urging the man to step just a tad closer, just to see… while the negative character is anticipating rejection in any event and deliberately steps on the girls' toes. Embarrassment is all around even with surrounding crowd because somewhere in the back they in the crowd all no what happened and that brings the embarrassment to their door. Every action on all accounts are very anticipated and pre-arranged. We use the name of an unfortunate incident to file this under.

It's about calling your wife but using the neighbours' wife's name. It is knowing you have to cut the lawn but the drowsiness just will not let go so you sit down and close your eyes for a second, just to relax for a second. The one character shifts one position while you were not attending procedure and you miss the opportunity to mow the lawn for one more week. This goes by the name of slipping the tong and nodding off.

Very typical of this is the driving absentmindedness. How many times did we sit back after arriving at our destination and thought how on earth did I pass this or that town? While driving your car it becomes as routinely as breathing so there is little to be positive with and less to be negative about. That is the chance the neutral character has been waiting for all week and he takes charge by letting your mind wonder to many destinations but the one you are heading to. Little harm can come from this under the normal, but when danger suddenly strikes you are gone, your positive is gone, you negative with all the adrenalin is completely absent and your neutral being neutral does not bother in any case. By the time all the absentees arrive at the scene just that second later, a horrible accident is in progress. This we call driver fatigue. The process lingers on while we are absent minded, not totally in control of the moment and shit happens. But shit happens because we want it to happen for that bit of excitement we do not need, but the characters do because they make misery, To them it is a case to change the situation to something more exciting, more emotional, to press our social standing or just to ease boredom. This goes by the name of the mind is wondering.

Someone may say something out of the order and a rage follows. Why would a rage follow when you know very well that what the other person said was unfounded and if that person does believe what he said he is so misinformed you should not even listen to such nonsense? But confrontation is about to blow like a rocket on a launch pad. You will show him what he said was untrue and he will take it back or pay with his life. Once again we allow the characters to shift and completely obstruct our normality. The best is that with you allowing the shifting you will press old issues long ago forgotten but the characters suddenly helped you remembering this or that and this is the last straw! Under the normal the previous incident hold such minor importance you would never remember about it but at that moment it does not even strike you as odd remembering such trivially while not surprising at all it is clear in your memory that moment. You feel you can murder while never in your life did you ever have a thought about how it may feel taking another person's life. Everything you experience is out of the ordinary and out of order yet it seems to you at that moment as being as normal as discussing the papers. Does your characters enjoy the excitement and taking you along for the ride. They take control and you take the mess afterwards not knowing how you will ever show your face again after such an incident. This we call losing your temper.

You walk down the street with your best suit on feeling as chirpy as a robin in mating season. Then suddenly your foot misses the curb as you were walking and you land face down in a crowd of people you never saw previously or know any one around you. There is not the slightest chance of you meeting any one of the spectators ever again. Yet your world plummets down the deepest mine shaft. You cannot see the light of day ever rising again. You hang your head in shame while trying to conquer the incredible urge to run from the scene as fast as you can. Not for a moment do you stop to think that it is not that shameful and happens to everybody many time throughout they're lives. That the people you hold so important can never be important because you will never see any of them ever again. Again you characters turned a situation around as they turned on you. This we call embarrassment.

All of us meet the challenge in battle without ever realising. A young man newly wed has a friend of his wife staying with the couple for a month or two, just until she can find some other accommodation. This young man is going ballistic with the fight. He is in rage about this beauty sharing a roof with him and with him working shifts he finds himself alone in the company of this woman because she is still in the market for employment. She page thorough many papers per day to find a suitable position but that does not take all day and with his wife being at her work there are many hours pleasuring about unchecked. Now he goes in spin. His positive character tells him of what he has in his wife and that he should not endanger his fragile and young marriage, his neutral tells him his wife is at work and need not to know while his negative went positive by telling him how beautiful this young female is. This we use the name temptation to identify. It is the clearest way we know of identifying the threesome.

We were all at one or other stage young in our lives and young at heart and know the feeling when you are running the hundred meters or you are on the Rugby field and there is that special person looking from the spectators end. You feel her eyes burn in your back and the excitement blows your senses. It is like someone ells takes control and you become that much better to the degree you cannot believe yourself, or you have gone pinching fruit (a favourite pass time amongst the Afrikaner up and until I was a teenager. That was before money and greed came in and made kids criminals for helping themselves to fruit at night). The other side of the coin was that you knew who ever gets hold of you, will tear of the skin from your behind, but that was part of the fun being the challenge to the danger. Never was police involved one way or the other and even coming home safely did not mean security because if your parents catch you, you have to go back to whom fruit you pinched belongs and fetch your licking. I was a hundred meter athlete in my time and was quite quick, but when caught in the act I saw these real slow guise and can-not-run mates in crime pass me and if Carl Lewis was there they will pass him as well. They beat me by a hundred meters on the three hundred meters and cross one and a half meter fences as if it was hurdles on the track. The neutral character was on guard all-night and got the other two pumped but on stand by. The moment surprise takes over the positive gets negative but charging past the negative to get the body in motion. The negative sees that being negative is a splendid way of getting away from the danger and re-passes the positive while the neutral is pushing both to get out of the way so that the negative can take command of the muscles. With in less than a heartbeat there are four characters (including yourself) that take control of muscles and there can be no fitter and more potent athlete occupying the body than the fear factor presenting the next few minutes. This we call motivation.

Then there is the young person that is the one-week in the dump because he cannot see what the world is all about. His girlfriend dropped him, he is in the middle if an exam he cannot see how he will ever pass and to top it all is the fact that it has been raining for three days while he has this camping trip planned. The positive character finds a few days rest while laying low in anticipation of the coming holiday forcing the negative character to take charge of his outlook on life while his neutral character takes charge of the schoolwork making him take his vacation early while exams are pressing. The next week he finds himself being over positive bout life, about his happiness and about all the opportunities in life that are waiting on him. The exams are history, he is vacating and met the girl of his dreams and have a song in the heart. To tell the world, he is the man of the moment and to prove that he has long uncombed hair, walks with a swing and doing his thing just to annoy his square parents that are living they're life in history. That is the neutral character in charge and being all-positive for one week/ month and then negative for one week/month does not surprise any one. The name we attach to this is growing up. It is all about finding the characters, meeting the characters and blending with the characters in order to fine tune for preparedness for the fight ahead.

We name the events. We know the events. We suffer through such events but never stop to think why it is taking place. Why would your mind start to wonder? If you are in control and you are in that position you should not find yourself out of control or somewhere ells while you were being there. If you were your body why would the mind go absent? If you were your body and mind why would you stray or over react in anger, pity, shame or stupidity. If only being in the body you are in as the atheists believe, then you should be in the body, because where ells will you be but in the body you are. The fight is on and you have to recognise your opponent because your opponent is you in person. Your opponent knows your weaknesses as you know them because you are your weaknesses. Your life is your fight and it is on for the rest of you natural life.

It is as common in every one as miss-placing keys, spilling milk, forgetting some one's name and such minor incidents bringing great embarrassment at the time but is as normal as breathing. It is no big psychological problem forming in the dark side of the sole where the brave does not dare. Yet it can be there. When ever a person comes up with a brilliant remark astonishing the person that made the remark much more that the person to whom the remark was directed is an example of the trio. When trying to repeat the brilliance the very next time we become the stuttering idiot that wishes the floor would dissolve the human body. We find an inability to repeat such brilliance. We all have astonished ourselves from time to time as much as we embarrass our selves from time to time, so there is no exclusion and only inclusion.

It is the person being unwilling to do a task and finding himself repeating an error that he knows he should not repeat but something is driving the person to repast his actions. The more he repeats the error the more he gets annoyed with himself but getting annoyed with yourself must be the most unaccomplished task you can accomplish. His normal reaction would be then to become annoyed with any object or person he may find displeasing and although the object or person has little to nothing to do with his actions it proves the best way to blow off steam. This I refer to as the boss syndrome because that is one of the perks a boss seems to have. Being negative about the task makes your mind go wondering to more pleasant places to be. By your negativity you are suppressing the positive character and that puts the neutral character in charge. But the neutral character has as little interest in you're being there and takes you away on more pleasant day dreaming trips to where you would rather be and that leaves the negative character to be in charge. Who is doing the task with every one gone…it is the negative character, and an unpleasant one at that being all alone and hell is on its way!

The person starts making deliberate errors that can be avoided but with his lack of enthusiasm he becomes absent-minded and that is when the accidents stars cropping up. It could be small annoying things but also it can be very serious injury coming from such little absent mindedness. The negative character wishes to draw all three characters attention to the fact that the negative character is as displeased with the situation and wants to be relived of the duty. The LAPSE in CONCENTRATION is no lapse in concentration and the disaster following (big or small) is as deliberate as the person slipping away on his daydreaming trip. The whole affair is one big disaster waiting to happen and always does happen. All parties involved claims innocence and protest to any involvement accept the negative character that now is relieved of his duties. The net result is that the negative character is the only winner in the end. Where does it come from? It comes from being human. What encourages the characters to become more dominant in some than in others are the better choice of question? There are at least a thousand possibilities I presume but it is more than just likely that it may present itself as a manifestation of a chemical imbalance in the brain. That does not explain the fact that it is present but may support the fact of more dominance in some individuals than in others. Being alive is about chemicals but the chemicals allow conducting electricity and are not life itself. The chemicals are a conductor but we all know the conductor is not electricity. But when the conductor goes hay-why the electricity goes hay-why. In that sense the chemicals will play a part but not play the part. When there are cross over of wire connections some life will flow in a direction where it actually is intended to be at other outlets. From what I have seen with some of my brain injured friends the control becomes absent but that does not mean the injury cause the characters because I have witnessed the characters in every person I have met. It is as much part of our personality s it is our personality.

The chemicals could be one aspect but underlying fears are the predominant issues that renders the characters the possibility of going out of control. An unhappy childhood brings an unhappy life. When the child is growing into a skew adult the adult will bring about another skew child. The main drive behind man is fear. Where fear is absent the man is a danger to him and to society. Fear brings about a conscience. The fear I refer to is not anxiety or being scared but having respect. Having respect for one's parents or teachers or the community and respect for the law. Above all the fear of God brings respect for God and that forms the basis for being man. Persons with personality problems and character flaws may have too much fear or none at all. It is a balance and it is the balance that keeps us upright. Underlying in the balance is respect for yourself and that you have with the respect you have for your parents. When being a child your parents are a substitute for God because only they can bring values and norms that will one day bring about a balanced member of society. Money cannot buy that.

I know that in itself having a conscious and adhering to the conscious does not say much because the alcoholic hides his booze from himself because his characters are in conflict. He knows he is on the road of self-destruction but normally that does not bother an alcoholic that much as they love drink more than life. It normally is the fact that the alcoholic knows about the destruction in his children and his love for his dear ones. He wishes to protect them from him but he has this massive problem that is stronger than his urge for life. To find a way to solve the problem he starts to hide the liquor from himself and in that way he stars to lie to himself. The characters are accusing him as they are destroying him and the escape is the destruction. When he is drunk he has no control and when he is sober he has no control. The characters are telling him he has no problem as much as they are accusing him about his problem and denial is also admitting where admitting then becomes blame and the blame he carries are more than he can carry therefore he drinks to escape the blame of his guilt.

It is not only liquor but also it is sex, drugs, pornography and gambling. Those are the weaknesses that puts people in the same situation as that of madness. It is living the life of lust knowing that that life of choice is what you choose and choosing such a life is the equivalent of destruction but modern society makes fighting thereof much harder than ever before and capitulating as easy as breathing. The guilt, fear, anger and despair comes in waves and in conflict where the alcoholic is in the middle with his problem out of control and out of his hands. His characters took charge and they destroy him as well as those surrounding him... all that loves and care for him but the biggest destruction is his destroying of the ones he love. Seeing the suffering in the lives of those poor, poor soles make me great fill for the small load I received and the ease with which I have to carry my small burden. In that way we're in a fight but the fight is all in private and we may acknowledge failure where there are great success as we may judge success where there is great failure. We on the outside judge what we see on the outside while the fight is on the inside where we can never see. A rehabilitated alcoholic must be a far bigger success than a successful achiever born with the golden spoon in the mouth but skins the cat in the dark by indulging in cravings of the night, all in the quiet his money can bay. It is about what you fight and not the way in which you fight or how you fight.

I am no philosopher of any sorts but thought about success and failure and my position in life where I as a person will never reach great achievements therefore the question I asked myself is will I die a failure because of that? My success is my children I leave behind and the success they may be. Not in great achievement because in the end even King Solomon declared it was all about chasing the wind. Leaving behind riches in money can bring as much despair as leaving behind poverty. The only way I may achieve success is leaving behind children that are of a better fabric than I was. The next generation must be better equipped, better evolved and better humans than was the previous generation to ensure evolution. What is taking place to my mind at present is devolution through out the western world. I do not have to bring proof about that because reading the paper or looking at nightlife will bring proof to any one wishing to see proof. How do we western man raise our future and how do we equip our future? Western society removed discipline from schools with admirable success. We removed the authority from the teachers as best we could. The teachers are through out the western world the lowest paid professionals in society. Being the lowest paid brings about the under achievers of society and any successful teacher leaves the profession for better pay in the private sector. This is the cancer of the new age we are facing. We all know my last remark is the truth and by that the parents wishes to compensate, but to modern man compensation is about money and paying to get rid of the guilt. Every nation culture community and individual admits that money is no measure for success and yet that is the norm we live by. Every person living a life is but a building block to establish the next generation. Man is an endangered species and animals are overpopulating the world and walking on two legs is not the criteria for human classification. The position you hold distancing yourself in thought and behaviour from the norms that animals uphold makes man or beast.

Animals are not in a struggle with their improving of the mental but in a struggle with the survival of the fittest, the one that can kill the best, run the fastest outsmart all others. That is not man. Man is about his fight to better his life and not his body or position. Man is about fighting the battle of the best in man. That is not modern man's ambitions. Modern man has ambitions making him the best animal on the planet because atheism propagates the fact that we are all members of the animal Kingdom. If man is that blind it may be best if man returns to the animal Kingdom where he thinks he belong. At least the loss will be small but smaller will be the gain of the animal world.

I wish to take you back to the indicial verse in that there were three things man would obtain when rating from the tree, and not only two...it is the three that every preacher misses ...**There was the tree...the tree of knowledge...knowledge of good, and knowledge of bad. You determine your knowledge...of good...of bad but above all it is your knowledge. The Bible does not name what is good or bad...that is your choice you make and that is your price you pay in the end...because the knowledge you accumulated will stand you either good or bad, as animal or man.**

It is not what you may accomplish or accumulate that has importance but what you learn through being man and standing apart from beast, that is what you take with.

Accept the following or reject the following but consider the following.

Because there's mathematically no nothing there must be a God.

By excluding nothing you have to include God because if there was nothing there was some scope for God being nothing and therefore non existing but since nothing is the only excluded number in mathematics you have to include God as a factor. What ever you dare to prove or disprove you can only prove or disprove through the human thought and understanding of concepts only human concept can understand.

I do realise with my statement about nothing and God it is stretching matters beyond the argument. When taking the argument that far the argument can include the existing of fairies and other fantasies, that is quite true, but by my placing a fairy and other fables in the realms of insignificance brings only harm to what remained of the child in me and I can assure you there is not much left in that sense. What ever is there is also in the infinity of my childhood memory. Not believing much in the fabric of the fantasy does not affect my position relating to the animal as the animal has little regard in such matters. Fantasy may have significant when I connect ethics to it as the Roman Catholics do with saints, demons, angels and such. When one can disregard such fantasies without affecting norms and values of the civilised, not harming the moral fibre of society there is little harm done by such removal. But when removing such norms for the enjoyment of being superior while the correctness of the argument in its core is invalid and, the structure of society goes to hell, a lot of question marks appear about the morals behind the motivation as to the motive behind the act. Man is moreover morals than life.

All life will destroy other life to its own benefit. A virus will kill his host for the benefit of one life cycle and at that a virus life cycle. What waste we may think. Killing a host only to spurn seems a lot of waste. Not so, because ticks will devour a cow alive without feeling any remorse. The same apply to a lion killing its pray. If there were no chance of the antelope finding means to escape the lion would start eating without killing saving itself the effort. It does not kill quickly through pity like humans do. It kills because of self-interest alone. When one of the herd falls victim to a lion attack the rest will start grazing thanking in that manner the victim for securing their position for one more day. There is no compassion, just self-centred egoistic drive to self-protection. When a person jumps into flames in an effort to save another person he is a hero. He receives praise from all concerned and may even land a medal for his effort. Why would we humans consider that as brave, being exceptional and above average?

When a person buys a car worth millions (in South African Mickey Mouse Money), because his business is going grate and he can afford to, but has his sister and his brother in law is working for him at minimum wage, not making ends meet in providing for their school going children, battling every day to put food on the table while he is making money like water flowing every one admires him for he is rich. He is treated very softly and no one dares to stand up to such a person because he has money. Such a person as a human is wasting breath because he may walk on two, but he may as well walk on four for the humanity he is. Still society regards that animal as a pillar of the community because he has money and he is a "sharp and intellectual business man". He should be shot at dawn for impersonating man whilst being beast. His positive has only one aim and that is to better his financial position securing a better admiration in society and establishing a front with him being god to the rest he sees as lesser mortals. His neutral sees his money drive as security and that satisfy while his negative is bent on destroying others because his positive is telling him how great full his sister and her "useless" husband should be for his generosity of providing food on their plates. Not once for the shortest and briefest instant will he ever give a thought that it is precisely the other way around. By their self-denying they are enriching him, but through his ego madness that thought never comes to mind. Then the philosophical will say that is man.

From our cosmic position we are in the centre of the universe. Where you may sit or stand is the very centre of the universe (your universe) because you can only relate to the universe having your individual singularity and where all other aspects in the cosmos are pointing away from your singularity to all other positions. We will always be 56 in a universe of 112. The earth will always be Π^2 in relation to the sun with the speed of light being $3\Pi^2$ from our position. It is not surprising we see ourselves on the edge of the Milky Way. From where we are we will be on the outer edge of our galactica. To the inside everything will be brighter and to the outside everything will be darker. That is a fact, but not a reality. The fact we may appreciate but the reality we can never understand. We are egocentric maniacs coming from the position of our singularity. That is the animal, the trio breathing our air. That is what the fight is about and what one may take with after death changes dimensions occupied. That is wisdom standing apart from knowledge and the animal has knowledge in intelligence but man has wisdom in extelegence. That parts man from beast. Man has to part from singularity's approach.

Us humans must fight to find our place outside singularity, outside self-interest and the strive to better of our position including finance. We have to relate from a position in divinity and not singularity to understand the cosmos and to understand life. If not we may well die as animals, and such possibility is there. If one person spent a lifetime killing his human in him, he may end as the animal he always strived to be. Being artist's engineers or preachers have nothing to do with it. Knowing the Bible or not has nothing to do with it. Being a believer or not has nothing to do with it. It is your approach to the cosmos that makes you part of the cosmos or that dimension above the cosmos. Accepting that one is not part of the cosmos but part of life in a cosmos puts one above the cosmos, but still in the cosmos. Never lose reality but accept responsibility. Persons believing in fairies and fantasies have the problem of over acclimating the positive while the neutral protects sanity with much of it going in the direction of a lost cause leaving the negative happy for destruction can only follow such obscurity. The art and artistic are typical in this. This is the line the drug addict follows to the last letter. By pumping the acid the positive hallucinates, the neutral exclude the incorrectness and the negative character are in seventh heaven as destruction is deliberate and decisive. The very same apply to a soldier in war where the positive character enjoys the killing as much as the negative character enjoys the being killed and the neutral character wishes the cruelty on them before they are upon us. When the soldier arrives back in society, society understands him as little as he understands himself because the fabric of human principles has gone skew. His characters, including him is as mixed as a milkshake fruit salad.

There is the middle, a precise middle where I as a person hold my personal relevancy as I see myself. What ever I attach or detach puts me in relevancy to others in life. That is the purpose of my fight with my characters. My precise middle must be very straight and when the middle leans toward any side in particular my middle is out of alignment. To secure a centre there must always be room for others and their opinion. When I say there is no God I go eccentric and when I say there is only God I go eccentric.

As it is in mathematics in the matter of the line and the dot, it is a question of you deciding the relevancy. It is your choice and only your choice to what relevancy you wish to place God as a presence or a factor. By declaring the absence of God your relevancy might reduce God to infinity but in your denial you place relevancy therefore relevancy remain be it in the infinity. It is to the peril of the denier that such a person excludes God as a factor because by placing God at a point of infinity next to zero the denier places himself next to the animal that excludes God because God excluded the animal from extelegence. Such a thought proves the fool. God allowed the animal excluding the admitting of a God because God exclude the animal from the dimension of being human therefore liability to questionability. In the perception lies the norm.

On the other side of the spectrum is the religiosity maniac and such is his idiocy he gives God the role of the animal being on call by prayer to serve the master's call. In all such cases the maniac places himself in eternity by creating a spot in the centre of eternity for himself providing him endless power having God as his slave or animal on a leach in acclimation to his eternal status he then places God just outside eternal big for that spot he reserved for himself. Through his effort by prayer he can heal, bless and doom…and God will obey as instructed. Such a fool proves the thought.

Through your relevancy with God you become the God that makes you your own God as much as you're own devil and your own forgiver as much as your own accuser. You make the relevancy and the relevancy proves you. Being positive is loathing and being negative is damning and being neutral is obstinate and in everything you confirm the applying of norms be they right or wrong, good or bad as it is the free choice you make proving your distance you have as a factor claiming life and location as a being from the animal. In the case of the animal the ride is free for the animal knows no better but man does and is liable through conscience placing relevancy to deeds. On the physical aspect in the being named man, the human body is an accumulation of singularity positioning divinity as one comprising of three identities without ever separating the trinity. The one holds the straight line the other forms the triangle and the third is the dome, the inclusive sphere, and the container having seven sides to the entire outside world. You cannot be in two sides of the universe simultaneously but you may observe both sides by being centre.

CLAIMING A POSITION IN SINGULARITY'S INFINITY IS LIFE'S DIVINITY. YOU LEARN TO LIVE AS MUCH AS YOU LIVE TO LEARN

Singularity has three in parts as does divinity have three in places. On the other side of the divide is the divine and as things are in the one side of the divide in total relation to the other side because all relevancies align as much as match in equilibrium and in as much as it is the cosmos forming one half of the divide it has to be the divine duplicating to establish equilibrium. **The purpose of man in life is to know your other entities, to learn to live in recognising the other entities, to take from them strength they can give but also to detach from their weaknesses, and above everything else recognise yourself in what you do as pure or spoilt. Gain knowledge the knowledge you have in the good you can acquire and the knowledge in the bad you have to detest. Then only you may serve a life of gain. Did I prove anything, well you are my judge! As I am the one promoting all things connect, so I am the man believing all things connect.**

Without a zero there cannot be death. Without a nothing there cannot be no God. Accept one and you cannot except one.

If it is that simple why is it complicated.

BEST WISHES,

PETRUS. (PEET) S. J. SCHUTTE

Relevant applying literature Go to Google Amazon.com: Peet Schutte: Books
http://www.amazon.com/s?ie=UTF8&page=1&rh=n%3A283155%2Cp_27%3APeet%20Schutte.
Oxford dictionary of Astronomy web site naturescosmicconcept

The Following books are all available from CreateSpace web site.
The Absolute Relevance of Singularity The Journal
The Absolute Relevance of Singularity The Unpublished Article
The Absolute Relevance of Singularity The Dissertation
The Absolute Relevance of Singularity in terms of Newton Book 0
The Absolute Relevance of Singularity in terms of Cosmic Physics Book 1
The Absolute Relevance of Singularity in terms of The Sound Barrier Book 2
The Absolute Relevance of Singularity in terms of The Four Cosmic Phenomena Book 3
The Absolute Relevance of Singularity in terms of The Cosmic Code Book 4
The Absolute Relevance of Singularity in terms of Life Book 5
The Absolute Relevance of Singularity in terms of Investigating Kepler Book 6
The Absolute Relevance of Singularity in terms of The Thesis Book 7
The Absolute Relevance of Singularity in terms of The Cosmic Creation Book 8

peet@naturescosmicconcept.co.za   mail.naturescosmicconcept.co.za

www.ingramcontent.com/pod-product-compliance
Lightning Source LLC
Chambersburg PA
CBHW080656190526

45169CB00006B/2145